An Introduction to Civil Engineering

An Introduction to Civil Engineering

Edited by
Archer Green

Larsen & Keller
www.larsen-keller.com

An Introduction to Civil Engineering
Edited by Archer Green
ISBN: 978-1-63549-006-0 (Hardback)

Larsen & Keller
Published by Larsen and Keller Education,
5 Penn Plaza,
19th Floor,
New York, NY 10001, USA

Cataloging-in-Publication Data

An introduction to civil engineering / edited by Archer Green.
p. cm.
Includes bibliographical references and index.
ISBN 978-1-63549-006-0
1. Civil engineering--Textbooks. I. Green, Archer.
TA147 .I58 2017
624--dc23

Printed and bound in China.

For more information regarding Larsen and Keller Education and its products, please visit the publisher's website www.larsen-keller.com

Table of Contents

Preface

Civil engineering is one of the most widely used engineering disciplines. It plays a crucial role in the planning of any city or village. This book unfolds the innovative aspects of civil engineering which will be crucial for the understanding of the field. The sub-disciplines, applications and various other aspects of civil engineering have been lucidly covered in this book. It also provides an overview of the relationship between civil engineering and urban planning. For someone with an interest and eye for detail, this book covers the most significant topics related to this subject.

A detailed account of the significant topics covered in this book is provided below:

Chapter 1- This chapter focuses on civil engineering. It is written to suit the students' need and provide them a good basic knowledge about the subject. It will discuss briefly the various sub-fields of civil engineering, building the base for the next detailed chapters.

Chapter 2- This chapter elaborates the different sub-fields of civil engineering. Some of the topics discussed are architectural, construction, earthquake, structural and geotechnical engineering. It will provide details about all the sub-fields and discuss their importance in civil engineering.

Chapter 3- Civil engineering is not only applied for environmental protection or social welfare. It has a plethora of commercial applications as well. This chapter will focus on such applications of civil engineering. The topics included in this chapter are transportation engineering, coastal management, construction surveying, dry dock, etc.

Chapter 4- Sustainable civil engineering is the need of the hour because of growing pollution. This chapter will discuss few of the topics which fall under the branch of sustainable civil engineering. The topics included are green building, dams, seawall, rain garden, land reclamation and permeable paving. This area is an important part of civil engineering and also plays a crucial role in environment conservation.

Chapter 5- One of the main applications of civil engineering is its vital importance in urban planning and development. Civil engineering is required for the development of any city or a village. This chapter will particularly discuss the topics like planned community, parks, shopping malls, municipal and urban engineering and urban planning. It is one of the most important roles civil engineering plays in our society. Therefore, this chapter is written in such a way that it provides fair information about this aspect of civil engineering to students.

I would like to make a special mention of my publisher who considered me worthy of this opportunity and also supported me throughout the process. I would also like to thank the editing team at the back-end who extended their help whenever required.

Editor

1

Introduction to Civil Engineering

This chapter focuses on civil engineering. It is written to suit the students' need and provide them a good basic knowledge about the subject. It will discuss briefly the various sub-fields of civil engineering, building the base for the next detailed chapters.

Philadelphia City Hall in the US is still the world's tallest masonry load-bearing structure.

Civil engineering is a professional engineering discipline that deals with the design, construction, and maintenance of the physical and naturally built environment, including works like roads, bridges, canals, dams, and buildings. Civil engineering is the second-oldest engineering discipline after military engineering, and it is defined to distinguish non-military engineering from military engineering. It is traditionally broken into several sub-disciplines including architectural engineering, environmental engineering, geotechnical engineering, control engineering, structural engineering, earthquake engineering, transportation engineering, forensic engineering, municipal or urban engineering, water resources engineering, materials engineering, wastewater engineering, offshore engineering, facade engineering, coastal engineering, construction surveying, and construction engineering. Civil engineering takes place in the public sector from municipal through to national governments, and in the private sector from individual homeowners through to international companies.

A multi-level stack interchange, buildings, houses, and park in Shanghai, China.

History of the Civil Engineering Profession

Leonhard Euler developed the theory explaining the buckling of columns

Engineering has been an aspect of life since the beginnings of human existence. The earliest practice of civil engineering may have commenced between 4000 and 2000 BC in Ancient Egypt, the Indus Valley Civilization, and Mesopotamia (Ancient Iraq) when humans started to abandon a nomadic existence, creating a need for the construction of shelter. During this time, transportation became increasingly important leading to the development of the wheel and sailing.

Until modern times there was no clear distinction between civil engineering and architecture, and the term engineer and architect were mainly geographical variations referring to the same occupation, and often used interchangeably. The construction of pyramids in Egypt (circa 2700–2500 BC) were some of the first instances of large structure constructions. Other ancient historic civil engineering constructions include the Qanat water management system (the oldest is older than 3000 years and longer than 71 km,) the Parthenon by Iktinos in Ancient Greece (447–438 BC), the Appian Way by Roman engineers (c. 312 BC), the Great Wall of China by General Meng T'ien under orders from Ch'in Emperor Shih Huang Ti (c. 220 BC) and the stupas constructed in ancient Sri Lanka like the Jetavanaramaya and the extensive irrigation works in Anuradhapura. The Romans developed civil structures throughout their empire, including especially aqueducts, insulae, harbors, bridges, dams and roads.

John Smeaton, the "father of civil engineering"

In the 18th century, the term civil engineering was coined to incorporate all things civilian as opposed to military engineering. The first self-proclaimed civil engineer was John Smeaton, who constructed the Eddystone Lighthouse. In 1771 Smeaton and some of his colleagues formed the Smeatonian Society of Civil Engineers, a group of leaders of the profession who met informally over dinner. Though there was evidence of some technical meetings, it was little more than a social society.

In 1818 the Institution of Civil Engineers was founded in London, and in 1820 the eminent engineer Thomas Telford became its first president. The institution received a Royal Charter in 1828, formally recognising civil engineering as a profession. Its charter defined civil engineering as:

the art of directing the great sources of power in nature for the use and convenience of man, as the means of production and of traffic in states, both for external and internal trade, as applied in the construction of roads, bridges, aqueducts, canals, river navigation and docks for internal intercourse and exchange, and in the construction of ports, harbours, moles, breakwaters and lighthouses, and in the art of navigation by artificial power for the purposes of commerce, and in the construction and application of machinery, and in the drainage of cities and towns.

History of Civil Engineering Education

The first private college to teach civil engineering in the United States was Norwich University, founded in 1819 by Captain Alden Partridge. The first degree in civil engineering in the United States was awarded by Rensselaer Polytechnic Institute in 1835. The first such degree to be awarded to a woman was granted by Cornell University to Nora Stanton Blatch in 1905.

In the UK during the early 19th century, the division between civil engineering and military engineering (served by the Royal Military Academy, Woolwich), coupled with the demands of the Industrial Revolution, spawned new engineering education initiatives: the Royal Polytechnic Institution was founded in 1838, the private College for Civil Engineers in Putney was established in 1839, and the UK's first Chair of Engineering was established at the University of Glasgow in 1840.

History of Civil Engineering

Chichen Itza was a large pre-Columbian city in Mexico built by the Maya people of the Post Classic. The northeast column temple also covers a channel that funnels all the rainwater from the complex some 40 metres (130 ft) away to a rejollada, a former cenote.

Civil engineering is the application of physical and scientific principles for solving the problems of society, and its history is intricately linked to advances in understanding of physics and mathematics throughout history. Because civil engineering is a wide ranging profession, including several separate specialized sub-disciplines, its history is linked to knowledge of structures, materials science, geography, geology, soils, hydrology, environment, mechanics and other fields.

A Roman aqueduct [built circa 19 BC] near Pont du Gard, France

Throughout ancient and medieval history most architectural design and construction was carried out by artisans, such as stonemasons and carpenters, rising to the role of master builder. Knowledge was retained in guilds and seldom supplanted by advances. Structures, roads and infrastructure that existed were repetitive, and increases in scale were incremental.

One of the earliest examples of a scientific approach to physical and mathematical problems applicable to civil engineering is the work of Archimedes in the 3rd century BC, including Archimedes Principle, which underpins our understanding of buoyancy, and practical solutions such as Archimedes' screw. Brahmagupta, an Indian mathematician, used arithmetic in the 7th century AD, based on Hindu-Arabic numerals, for excavation (volume) computations.

The Civil Engineer

Education and Licensure

Surveying students with professor at the Helsinki University of Technology in the late 19th century.

Civil engineers typically possess an academic degree in civil engineering. The length of study is three to five years, and the completed degree is designated as a bachelor of engineering, or a bachelor of science. The curriculum generally includes classes in physics, mathematics, project management, design and specific topics in civil engineering. After taking basic courses in most sub-disciplines of civil enginecring, they move onto specialize in one or more sub-disciplines at advanced levels. While an undergraduate degree (BEng/BSc) normally provides successful students with industry-accredited qualification, some academic institutions offer post-graduate degrees (MEng/MSc), which allow students to further specialize in their particular area of interest.

In most countries, a bachelor's degree in engineering represents the first step towards professional certification, and a professional body certifies the degree program. After completing a certified degree program, the engineer must satisfy a range of requirements (including work experience and exam requirements) before being certified. Once certified, the engineer is designated as a professional engineer (in the United States, Canada and South Africa), a chartered engineer (in most Commonwealth countries), a chartered professional engineer (in Australia and New Zealand), or a European engineer (in most countries of the European Union). There are international agreements between relevant professional bodies to allow engineers to practice across national borders.

The benefits of certification vary depending upon location. For example, in the United States and Canada, "only a licensed professional engineer may prepare, sign and seal, and submit engineering plans and drawings to a public authority for approval, or seal engineering work for public and private clients." This requirement is enforced under provincial law such as the Engineers Act in Quebec.

No such legislation has been enacted in other countries including the United Kingdom. In Australia, state licensing of engineers is limited to the state of Queensland. Almost all certifying bodies maintain a code of ethics which all members must abide by.

Engineers must obey contract law in their contractual relationships with other parties. In cases where an engineer's work fails, he may be subject to the law of tort of negligence, and in extreme cases, criminal charges. An engineer's work must also comply with numerous other rules and regulations such as building codes and environmental law.

Sub-Disciplines

The Akashi Kaikyō Bridge in Japan, currently the world's longest suspension span.

In general, civil engineering is concerned with the overall interface of human created fixed projects with the greater world. General civil engineers work closely with surveyors and specialized civil engineers to design grading, drainage, pavement, water supply, sewer service,dams, electric and communications supply. General civil engineering is also referred to as site engineering, a branch of civil engineering that primarily focuses on converting a tract of land from one usage to another. Site engineers spend time visiting project sites, meeting with stakeholders, and preparing construction plans. Civil engineers apply the principles of geotechnical engineering, structural engineering, environmental engineering, transportation engineering and construction engineering to residential, commercial, industrial and public works projects of all sizes and levels of construction.

Materials Science and Engineering

Materials science is closely related to civil engineering. It studies fundamental characteristics of materials, and deals with ceramics such as concrete and mix asphalt concrete, strong metals such as aluminum and steel, and polymers including polymethylmethacrylate (PMMA) and carbon fibers.

Materials engineering involves protection and prevention (paints and finishes). Alloying combines two types of metals to produce another metal with desired properties. It incorporates elements of applied physics and chemistry. With recent media attention on nanoscience and nanotechnology, materials engineering has been at the forefront of academic research. It is also an important part of forensic engineering and failure analysis.

Coastal Engineering

Oosterscheldekering, a storm surge barrier in the Netherlands.

Coastal engineering is concerned with managing coastal areas. In some jurisdictions, the terms sea defense and coastal protection mean defense against flooding and erosion, respectively. The term coastal defense is the more traditional term, but coastal management has become more popular as the field has expanded to techniques that allow erosion to claim land.

Construction Engineering

Construction engineering involves planning and execution, transportation of materials, site development based on hydraulic, environmental, structural and geotechnical engineering. As construction firms tend to have higher business risk than other types of civil engineering firms do, construction engineers often engage in more business-like transactions, for example, drafting and reviewing contracts, evaluating logistical operations, and monitoring prices of supplies.

Earthquake Engineering

Earthquake engineering involves designing structures to withstand hazardous earthquake exposures. Earthquake engineering is a sub-discipline of structural engineering. The main objectives of earthquake engineering are to understand interaction of structures on the shaky ground; foresee the consequences of possible earthquakes; and design, construct and maintain structures to perform at earthquake in compliance with building codes.

Environmental Engineering

Water pollution

Environmental engineering is the contemporary term for sanitary engineering, though sanitary engineering traditionally had not included much of the hazardous waste management and environmental remediation work covered by environmental engineering. Public health engineering and environmental health engineering are other terms being used.

Environmental engineering deals with treatment of chemical, biological, or thermal wastes, purification of water and air, and remediation of contaminated sites after waste disposal or accidental contamination. Among the topics covered by environmental engineering are pollutant transport, water purification, waste water treatment, air pollution, solid waste treatment, and hazardous waste management. Environmental engineers administer pollution reduction, green engineering, and industrial ecology. Environmental engineers also compile information on environmental consequences of proposed actions.

Geotechnical Engineering

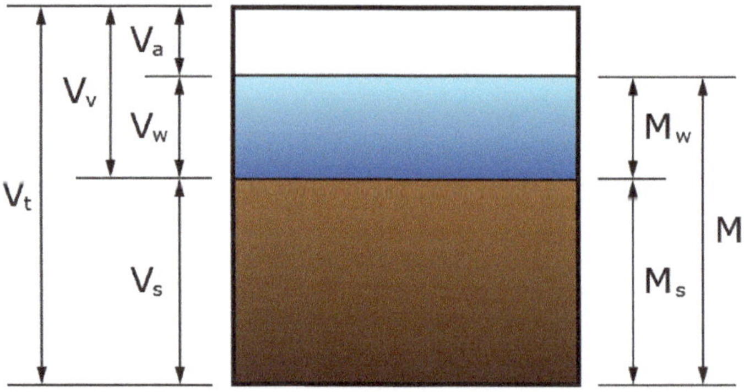

A phase diagram of soil indicating the weights and volumes of air, soil, water, and voids.

Geotechnical engineering studies rock and soil supporting civil engineering systems. Knowledge from the field of soil science, materials science, mechanics, and hydraulics is applied to safely and economically design foundations, retaining walls, and other structures. Environmental efforts to protect groundwater and safely maintain landfills have spawned a new area of research called geoenvironmental engineering.

Identification of soil properties presents challenges to geotechnical engineers. Boundary conditions are often well defined in other branches of civil engineering, but unlike steel or concrete, the material properties and behavior of soil are difficult to predict due to its variability and limitation on investigation. Furthermore, soil exhibits nonlinear (stress-dependent) strength, stiffness, and dilatancy (volume change associated with application of shear stress), making studying soil mechanics all the more difficult.

Water Resources Engineering

Hoover dam

Water resources engineering is concerned with the collection and management of water (as a natural resource). As a discipline it therefore combines elements of hydrology, environmental science, meteorology, conservation, and resource management. This area of civil engineering relates to the prediction and management of both the quality and the quantity of water in both underground (aquifers) and above ground (lakes, rivers, and streams) resources. Water resource engineers analyze and model very small to very large areas of the earth to predict the amount and content of water as it flows into, through, or out of a facility. Although the actual design of the facility may be left to other engineers.

Hydraulic engineering is concerned with the flow and conveyance of fluids, principally water. This area of civil engineering is intimately related to the design of pipelines, water supply network, drainage facilities (including bridges, dams, channels, culverts, levees, storm sewers), and canals. Hydraulic engineers design these facilities using the concepts of fluid pressure, fluid statics, fluid dynamics, and hydraulics, among others.

Structural Engineering

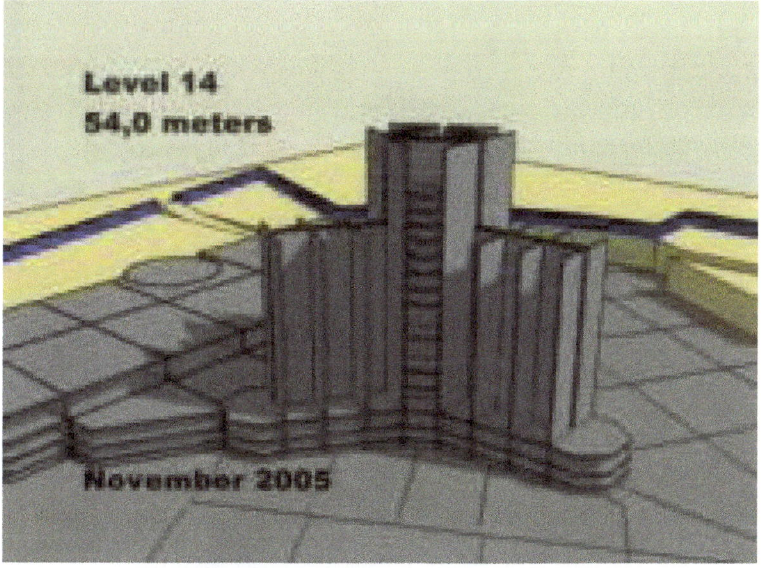

Pla Burj Khalifa animation of construction process

Structural engineering is concerned with the structural design and structural analysis of buildings, bridges, towers, flyovers (overpasses), tunnels, off shore structures like oil and gas fields in the sea, aerostructure and other structures. This involves identifying the loads which act upon a structure and the forces and stresses which arise within that structure due to those loads, and then designing the structure to successfully support and resist those loads. The loads can be self weight of the structures, other dead load, live loads, moving (wheel) load, wind load, earthquake load, load from temperature change etc.

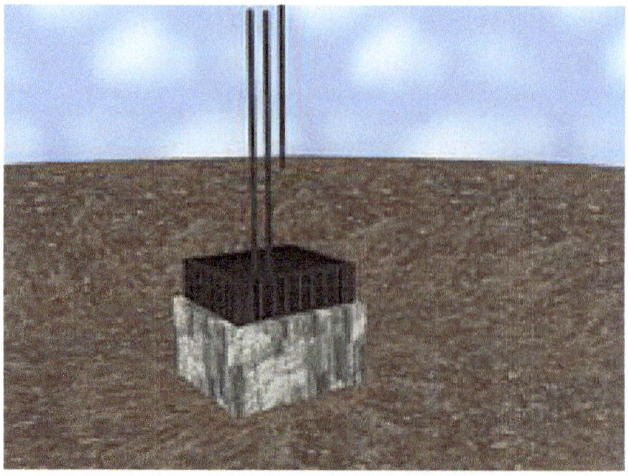

Play me Shallow foundation construction example

The structural engineer must design structures to be safe for their users and to successfully fulfill the function they are designed for (to be *serviceable*). Due to the nature of some loading conditions, sub-disciplines within structural engineering have emerged, including wind engineering and earthquake engineering.

Design considerations will include strength, stiffness, and stability of the structure when subjected to loads which may be static, such as furniture or self-weight, or dynamic, such as wind, seismic, crowd or vehicle loads, or transitory, such as temporary construction loads or impact. Other considerations include cost, constructability, safety, aesthetics and sustainability.

Surveying

Civil engineering student using a theodolite.

Surveying is the process by which a surveyor measures certain dimensions that occur on or near the surface of the Earth. Surveying equipment, such as levels and theodolites, are used for accurate measurement of angular deviation, horizontal, vertical and slope distances. With computerisation, electronic distance measurement (EDM), total stations, GPS surveying and laser scanning have to a large extent supplanted traditional instruments. Data collected by survey measurement is converted into a graphical representation of the Earth's surface in the form of a map. This information is then used by civil engineers, contractors and realtors to design from, build on, and trade, respectively. Elements of a structure must be sized and positioned in relation to each other and to site boundaries and adjacent structures. Although surveying is a distinct profession with separate qualifications and licensing arrangements, civil engineers are trained in the basics of surveying and mapping, as well as geographic information systems. Surveyors also lay out the routes of railways, tramway tracks, highways, roads, pipelines and streets as well as position other infrastructure, such as harbors, before construction.

Land Surveying

BLM cadastral survey marker from 1992 in San Xavier, Arizona.

In the United States, Canada, the United Kingdom and most Commonwealth countries land surveying is considered to be a separate and distinct profession. Land surveyors are not considered to be engineers, and have their own professional associations and licensing requirements. The services of a licensed land surveyor are generally required for boundary surveys (to establish the boundaries of a parcel using its legal description) and subdivision plans (a plot or map based on a survey of a parcel of land, with boundary lines drawn inside the larger parcel to indicate the creation of new boundary lines and roads), both of which are generally referred to as Cadastral surveying.

Construction Surveying

Construction surveying is generally performed by specialised technicians. Unlike land surveyors, the resulting plan does not have legal status. Construction surveyors perform the following tasks:

- Surveying existing conditions of the future work site, including topography, existing buildings and infrastructure, and underground infrastructure when possible;
- "lay-out" or "setting-out": placing reference points and markers that will guide the construction of new structures such as roads or buildings;
- Verifying the location of structures during construction;
- As-Built surveying: a survey conducted at the end of the construction project to verify that the work authorized was completed to the specifications set on plans.

Transportation Engineering

The engineering of this roundabout in Bristol, England, attempts to make traffic flow free-moving

Transportation engineering is concerned with moving people and goods efficiently, safely, and in a manner conducive to a vibrant community. This involves specifying, designing, constructing, and maintaining transportation infrastructure which includes streets, canals, highways, rail systems, airports, ports, and mass transit. It includes areas such as transportation design, transportation planning, traffic engineering, some aspects of urban engineering, queueing theory, pavement engineering, Intelligent Transportation System (ITS), and infrastructure management.

Forensic Engineering

Forensic engineering is the investigation of materials, products, structures or components that fail or do not operate or function as intended, causing personal injury or damage to property. The consequences of failure are dealt with by the law of product liability. The field also deals with retracing processes and procedures leading to accidents in operation of vehicles or machinery. The subject is applied most commonly in civil law cases, although it may be of use in criminal law cases. Generally the purpose of a Forensic engineering investigation is to locate cause or causes of failure with a view to improve performance or life of a component, or to assist a court in determining the facts of an accident. It can also involve investigation of intellectual property claims, especially patents.

Municipal or Urban Engineering

Lake Chapultepec

Municipal engineering is concerned with municipal infrastructure. This involves specifying, designing, constructing, and maintaining streets, sidewalks, water supply networks, sewers, street lighting, municipal solid waste management and disposal, storage depots for various bulk materials used for maintenance and public works (salt, sand, etc.), public parks and cycling infrastructure. In the case of underground utility networks, it may also include the civil portion (conduits and access chambers) of the local distribution networks of electrical and telecommunications services. It can also include the optimizing of waste collection and bus service networks. Some of these disciplines overlap with other civil engineering specialties, however municipal engineering focuses on the coordination of these infrastructure networks and services, as they are often built simultaneously, and managed by the same municipal authority. Municipal engineers may also design the site civil works for large buildings, industrial plants or campuses (i.e. access roads, parking lots, potable water supply, treatment or pretreatment of waste water, site drainage, etc.)

Control Engineering

Control engineering (or *control systems engineering*) is the branch of civil engineering discipline that applies control theory to design systems with desired behaviors. The practice uses sensors to measure the output performance of the device being controlled (often a vehicle) and those measurements can be used to give feedback to the input actuators that can make corrections toward desired performance. When a device is designed to perform without the need of human inputs for correction it is called automatic control (such as cruise control for regulating a car's speed). Multidisciplinary in nature, control systems engineering activities focus on implementation of control systems mainly derived by mathematical modeling of systems of a diverse range.

Civil Engineering Associations

The Falkirk Wheel in Scotland

- American Society of Civil Engineers
- Canadian Society for Civil Engineering
- Conselho Federal de Engenharia e Agronomia (pt)

- Earthquake Engineering Research Institute
- Engineers Australia
- European Federation of National Engineering Associations
- International Federation of Consulting Engineers
- Institution of Civil Engineers
- The Institution of Civil Engineering Surveyors
- Institution of Engineers (India)
- Institution of Engineers of Ireland
- Institute of Transportation Engineers
- Pakistan Engineering Council
- Philippine Institute of Civil Engineers
- Transportation Research Board

References

- *Oakes, William C.; Leone, Les L.; Gunn, Craig J. (2001). Engineering Your Future. Great Lakes Press. ISBN 1-881018-57-1.*
- p. 4 of *Mays, L. (30 August 2010). Ancient Water Technologies. Springer. ISBN 978-90-481-8631-0.*
- Chen W-F, Scawthorn C. Earthquake Engineering Handbook, CRC Press, 2003, ISBN 0-8493-0068-1, Chapter 2
- *"Singapore's Circle Line criminal trial started". New Civil Engineer. Retrieved 16 November 2013.*

2

Sub-disciplines of Civil Engineering

This chapter elaborates the different sub-fields of civil engineering. Some of the topics discussed are architectural, construction, earthquake, structural and geotechnical engineering. It will provide details about all the sub-fields and discuss their importance in civil engineering.

Architectural Engineering

César Pelli's Ratner Athletic Center uses cables and masts as load-bearing devices.

Architectural engineering, also known as building engineering, is the application of engineering principles and technology to building design and construction. Definitions of an architectural engineer may refer to:

- An engineer in the structural, mechanical, electrical, construction or other engineering fields of building design and construction.
- A licensed engineering professional in parts of the United States.
- Architectural engineers are those who work with other engineers and architects for the designing and construction of buildings.

Engineering for Building

Structural Engineering

Structural engineering involves the analysis and design of physical objects (buildings, bridges, equipment supports, towers and walls). Those concentrating on buildings are responsible for the structural performance of a large part of the built environment and are, sometimes, informally referred to as "building engineers". Structural engineers require expertise in strength of materials and in the seismic design of structures covered by earthquake engineering. Architectural Engineers sometimes structural as one aspect of their designs; the structural discipline when practiced as a specialty works closely with architects and other engineering specialists.

Mechanical, Electrical and Plumbing (Mep)

MEP room in a building

Mechanical engineering and electrical engineering engineers are specialists, commonly referred to as (mechanical, electrical, and plumbing) when engaged in the building design fields. Also known as "building services engineering" in the United Kingdom, Canada, and Australia. Mechanical engineers often design and oversee the heating, ventilation and air conditioning (HVAC), plumbing, and rain gutter systems. Plumbing designers often include design specifications for simple active fire protection systems, but for more complicated projects, fire protection engineers are often separately retained. Electrical engineers are responsible for the building's power distribution, telecommunication, fire alarm, signalization, lightning protection and control systems, as well as lighting systems.

The Architectural Engineer (Pe) in the United States

In many jurisdictions of the United States, the architectural engineer is a licensed engineering professional. Usually a graduate of an architectural engineering university program preparing students to perform whole-building design in competition with architect-engineer teams; or for practice in one of structural, mechanical or electrical fields of building design, but with an appreciation of integrated architectural requirements.

Formal architectural engineering education, following the engineering model of earlier disciplines, developed in the late 19th century, and became widespread in the United States by the mid-20th century. With the establishment of a specific "architectural engineering" NCEES Professional Engineering registration examination in the 1990s, and first offering in April 2003, architectural engineering became recognized as a distinct engineering discipline in the United States. Architectural engineers are not entitled to practice architecture unless they are also licensed as architects.

The Architect as Architectural Engineer

In some countries, the practice of architecture includes planning, designing and overseeing the building's construction, and architecture, as a profession providing architectural services, is referred to as "architectural engineering". In Japan, a "first-class architect" plays the dual role of architect and building engineer, although the services of a licensed "structural design first-class architect" are required for buildings over a certain scale.

In some languages, such as Korean and Arabic, "architect" is literally translated as "architectural engineer". In some countries, an "architectural engineer" (such as the *ingegnere edile* in Italy) is entitled to practice architecture and is often referred to as an architect. These individuals are often also structural engineers. In other countries, such as Germany, Austria, Iran, and most of the Arab countries, architecture graduates receive an engineering degree (*Dipl.-Ing. – Diplom-Ingenieur*).

In Spain, an "architect" has a technical university education and legal powers to carry out building structure and facility projects.

In Brazil, architects and engineers used to share the same accreditation process (CONFEA – Federal Council of Engineering, Architecture and Agronomy). Now the Brazilian architects and urbanists have their own accreditation process (CAU – Architecture and Urbanism Council). Besides traditional architecture design training, Brazilian architecture courses also offer complementary training in engineering disciplines such as structural, electrical, hydraulic and mechanical engineering. After graduation, architects can be fully responsible for design and construction in these areas (except in electric wiring, where the architect autonomy is limited to systems up to 30kVA), applied to buildings, urban environment, built cultural heritage, landscape planning, interiorscape planning and regional planning.

In Greece licensed architectural engineers are graduates from architecture faculties that belong to the Polytechnic University, obtaining an "Engineering Diploma". They graduate after 5 years of studies and are fully entitled architects once they become members of the Technical Chamber of Greece. The Engineering Diploma equals a master's degree in ECTS units (300) according to the Bologna Accords.

Education

The architectural, structural, mechanical and electrical engineering branches each have well established educational requirements that are usually fulfilled by completion of a university program.

An air handling unit is used for the heating and cooling of air in a central location (click on image for legend). Bringing together knowledge of acoustic engineering and HVAC is one example of the multi-disciplined nature of architectural engineering

Architectural Engineering as a Single Integrated Field of Study

What differentiates architectural engineering from architecture (architect) as a separate and single, integrated field of study, compared to other engineering disciplines, is its multi-disciplinary engineering approach. Through training in and appreciation of architecture, the field seeks integration of building systems within its overall building design. Architectural engineering includes the design of building systems including heating, ventilation and air conditioning (HVAC), plumbing, fire protection, electrical, lighting, architectural acoustics, and structural systems. In some university programs, students are required to concentrate on one of the systems; in others, they can receive a generalist architectural or building engineering degree.

Construction Engineering

Construction engineering technology is a professional discipline that deals with the designing, planning, construction, and management of infrastructures such as roads, tunnels, bridges, airports, railroads, facilities, buildings, dams, and utilities. These technicians are unique such that they are a cross between civil engineers and construction managers. Construction engineering technologists learn the designing aspect much like civil engineers and construction site management functions much like construction managers.

At the educational level, construction managers are not as focused on design work as they are on construction procedures, methods, and people management. Their primary concern is to deliver a project on time, within budget, and of the desired quality.

The difference between a construction engineering technologist and a civil engineer is that civil engineering is an engineering discipline. Civil engineering students concentrate more on the design work, which is more analytical, gearing them toward a career as a design professional. This essentially requires them to take a multitude challenging design courses. Construction engineering students take basic design courses as well as construction management courses. This allows them to understand both basic design functions as well as the building requirements needed to design and build today's infrastructures.

Work Activities

Depending on which career the construction engineer has chosen to follow, an entry-level design engineer normally provides support to project managers and assist with creating conceptual designs, scopes, and cost estimates for the planning and construction of approved projects. It should be noted that a career in design work does require a professional engineer license (PE). Individuals who pursue this career path are strongly advised to sit for the Engineer In Training exam (EIT) while in college as it takes five years (4 years in USA) post graduate to obtain the PE license.

Entry-level construction manager positions are typically called project engineers or assistant project engineers. They are responsible for preparing purchasing requisitions, processing change orders, preparing monthly budgeting reports, and handling meeting minutes. The construction management position does not necessarily require a PE license; however possessing one does make the individual more marketable, as the PE license allows the individual to sign off on temporary structure designs.

Abilities

Construction engineers are problem solvers. They contribute to the creation of infrastructure that best meets the unique demands of its environment. They must be able to understand infrastructure life cycles. When compared and contrasted to design engineers, construction engineers bring to the table their own unique perspectives for solving technical challenges with clarity and imagination. While individuals considering this career path should certainly have a strong understanding of mathematics and science, many other skills are also highly desirable, including critical and analytical thinking, time management, people management, and good communication skills.

Educational Requirements

Individuals looking to obtain a construction engineering degree must first ensure that the program is accredited by EAC or Technology Accreditation Commission (TAC) of the Accreditation Board for Engineering and Technology (ABET). ABET accreditation is assurance that a college or university program meets the quality standards established by the profession for which it prepares its students. In the US there are currently twenty-five programs that exist in the entire country so careful college consideration is advised.

A typical construction engineering curriculum is a mixture of engineering mechanics, engineering design, construction management and general science and mathematics. This usually leads to a Bachelor of Science degree. The B.S. degree along with some design or construction experience is sufficient for most entry level positions. Graduate schools may be an option for those who want to go further in depth of the construction and engineering subjects taught at the undergraduate level. In most cases construction engineering graduates look to either civil engineering, engineering management, or business administration as a possible graduate degree.

Job Prospects

Job prospects for construction engineers generally have a strong cyclical variation. For example, starting in 2008 - continuing until at least 2011 - job prospects have been poor due to the collapse of housing bubbles in many parts of the world. This sharply reduced demand for construction, forced construction professionals towards infrastructure construction and therefore increased the competition faced by established and new construction engineers. This increased competition, and a core reduction in quantity demand is in parallel with a possible shift in the demand for construction engineers due to the automation of many engineering tasks, overall resulting in reduced prospects for construction engineers. In early 2010 the United States construction industry had a 27% unemployment rate, this is nearly three times higher than the 9.7% national average unemployment rate. The construction unemployment rate (including tradesmen) is comparable to the United States 1933 unemployment rate - the lowest point of the Great Depression - of 25%.

Remuneration

The average salary for a civil engineer in the UK depends on the sector, and more specifically the level of experience of the individual. A 2010 survey of the remuneration and benefits of those occupying jobs in construction and the built environment industry showed that the average salary of a civil engineer in the UK is £29,582. In the United States, as of May 2013, the average was $85,640. The average salary varies depending on experience, for example the average annual salary for a civil engineer with between 3 and 6 years experience is £23,813. For those with between 14 and 20 years experience the average is £38,214.

Earthquake Engineering

Earthquake engineering is an interdisciplinary branch of engineering that designs and analyzes structures, such as buildings and bridges, with earthquakes in mind. Its overall goal is to make such structures more resistant to earthquakes. An earthquake (or seismic) engineer aims to construct structures that will not be damaged in minor shaking and will avoid serious damage or collapse in a major earthquake. Earthquake engineering is the scientific field concerned with protecting society, the natural environment, and the man-made environment from earthquakes by limiting the seismic risk to socio-economically acceptable levels

Shake-table crash testing of a regular building model (left) and a base-isolated building model (right) at UCSD

Traditionally, it has been narrowly defined as the study of the behavior of structures and geo-structures subject to seismic loading; it is considered as a subset of structural engineering, geotechnical engineering, mechanical engineering, chemical engineering, applied physics, etc. However, the tremendous costs experienced in recent earthquakes have led to an expansion of its scope to encompass disciplines from the wider field of civil engineering, mechanical engineering and from the social sciences, especially sociology, political science, economics and finance.

The main objectives of earthquake engineering are:

- Foresee the potential consequences of strong earthquakes on urban areas and civil infrastructure.
- Design, construct and maintain structures to perform at earthquake exposure up to the expectations and in compliance with building codes.

A properly engineered structure does not necessarily have to be extremely strong or expensive. It has to be properly designed to withstand the seismic effects while sustaining an acceptable level of damage.

Seismic Loading

Tokyo Skytree, equipped with a tuned mass damper, is the world's tallest tower and is the world's second tallest structure.

Seismic loading means application of an earthquake-generated excitation on a structure (or geo-structure). It happens at contact surfaces of a structure either with the ground, with adjacent structures, or with gravity waves from tsunami. The loading that is expected at a given location on the Earth's surface is estimated by engineering seismology. It is related to the seismic hazard of the location.

Seismic Performance

Earthquake or seismic performance defines a structure's ability to sustain its main functions, such as its safety and serviceability, *at* and *after* a particular earthquake exposure. A structure is normally considered *safe* if it does not endanger the lives and well-being of those in or around it by partially or completely collapsing. A structure may be considered *serviceable* if it is able to fulfill its operational functions for which it was designed.

Basic concepts of the earthquake engineering, implemented in the major building codes, assume that a building should survive a rare, very severe earthquake by sustaining significant damage but without globally collapsing. On the other hand, it should remain operational for more frequent, but less severe seismic events.

Seismic Performance Assessment

Engineers need to know the quantified level of the actual or anticipated seismic performance associated with the direct damage to an individual building subject to a specified ground shaking. Such an assessment may be performed either experimentally or analytically.

Experimental Assessment

Experimental evaluations are expensive tests that are typically done by placing a (scaled) model of the structure on a shake-table that simulates the earth shaking and observing its behavior. Such kinds of experiments were first performed more than a century ago. Only recently has it become possible to perform 1:1 scale testing on full structures.

Due to the costly nature of such tests, they tend to be used mainly for understanding the seismic behavior of structures, validating models and verifying analysis methods. Thus, once properly validated, computational models and numerical procedures tend to carry the major burden for the seismic performance assessment of structures.

Seismic performance assessment or seismic structural analysis is a powerful tool of earthquake engineering which utilizes detailed modelling of the structure together with methods of structural analysis to gain a better understanding of seismic performance of building and non-building structures. The technique as a formal concept is a relatively recent development.

In general, seismic structural analysis is based on the methods of structural dynamics. For decades, the most prominent instrument of seismic analysis has been the earthquake response spectrum method which also contributed to the proposed building code's concept of today.

Analytical/Numerical Assessment

Snapshot from shake-table video of a 6-story non-ductile concrete building destructive testing

However, such methods are good only for linear elastic systems, being largely unable to model the structural behavior when damage (i.e., non-linearity) appears. Numerical *step-by-step integration* proved to be a more effective method of analysis for multi-degree-of-freedom structural systems with significant non-linearity under a transient process of ground motion excitation.

Basically, numerical analysis is conducted in order to evaluate the seismic performance of buildings. Performance evaluations are generally carried out by using non-linear static pushover analysis or nonlinear time-history analysis. In such analyses, it is essential to achieve accurate non-linear modeling of structural components such as beams, columns, beam-column joints, shear walls etc. Thus, experimental results play an important role in determining the modeling parameters of individual components, especially those that are subject to significant non-linear deformations. The individual components are then assembled to create a full non-linear model of the structure. Thus created models are analyzed to evaluate the performance of buildings.

The capabilities of the structural analysis software are a major consideration in the above process as they restrict the possible component models, the analysis methods available and, most importantly, the numerical robustness. The latter becomes a major consideration for structures that venture into the non-linear range and approach global or local collapse as the numerical solution becomes increasingly unstable and thus difficult to reach. There are several commercially available Finite Element Analysis software's such as CSI-SAP2000 and CSI-PERFORM-3D and Scia Engineer-ECtools which can be used for the seismic performance evaluation of buildings. Moreover, there is research-based finite element analysis platforms such as OpenSees, RUAUMOKO and the older DRAIN-2D/3D, several of which are now open source.

Research for Earthquake Engineering

Shake-table testing of Friction Pendulum Bearings at EERC

Research for earthquake engineering means both field and analytical investigation or experimentation intended for discovery and scientific explanation of earthquake engineering related facts, revision of conventional concepts in the light of new findings, and practical application of the developed theories.

The National Science Foundation (NSF) is the main United States government agency that supports fundamental research and education in all fields of earthquake engineering. In particular, it focuses on experimental, analytical and computational research on design and performance enhancement of structural systems.

E-Defense Shake Table

The Earthquake Engineering Research Institute (EERI) is a leader in dissemination of earthquake engineering research related information both in the U.S. and globally.

A definitive list of earthquake engineering research related shaking tables around the world may be found in Experimental Facilities for Earthquake Engineering Simulation Worldwide. The most prominent of them is now E-Defense Shake Table in Japan.

Major U.S. Research Programs

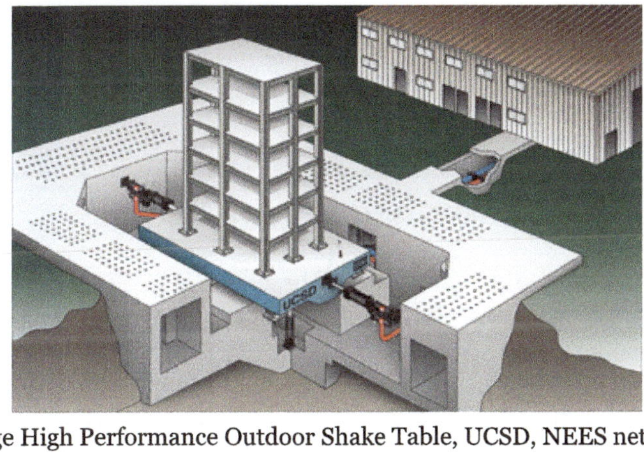

Large High Performance Outdoor Shake Table, UCSD, NEES network

NSF also supports the George E. Brown, Jr. Network for Earthquake Engineering Simulation

The NSF Hazard Mitigation and Structural Engineering program (HMSE) supports research on new technologies for improving the behavior and response of structural systems subject to earthquake hazards; fundamental research on safety and reliability of constructed systems; innovative developments in analysis and model based simulation of structural behavior and response including soil-structure interaction; design concepts that improve structure performance and flexibility; and application of new control techniques for structural systems.

(NEES) that advances knowledge discovery and innovation for earthquakes and tsunami loss reduction of the nation's civil infrastructure and new experimental simulation techniques and instrumentation.

The NEES network features 14 geographically-distributed, shared-use laboratories that support several types of experimental work: geotechnical centrifuge research, shake-table tests, large-scale structural testing, tsunami wave basin experiments, and field site research. Participating universities include: Cornell University; Lehigh University; Oregon State University; Rensselaer Polytechnic Institute; University at Buffalo, State University of New York; University of California, Berkeley; University of California, Davis; University of California, Los Angeles; University of California, San Diego; University of California, Santa Barbara; University of Illinois, Urbana-Champaign; University of Minnesota; University of Nevada, Reno; and the University of Texas, Austin.

NEES at Buffalo testing facility

The equipment sites (labs) and a central data repository are connected to the global earthquake engineering community via the NEEShub website. The NEES website is powered by HUBzero software developed at Purdue University for nanoHUB specifically to help the scientific community share resources and collaborate. The cyberinfrastructure, connected via Internet2, provides interactive simulation tools, a simulation tool development area, a curated central data repository, animated presentations, user support, telepresence, mechanism for uploading and sharing resources, and statistics about users and usage patterns.

This cyberinfrastructure allows researchers to: securely store, organize and share data within a standardized framework in a central location; remotely observe and participate in experiments through the use of synchronized real-time data and video; collaborate with colleagues to facilitate the planning, performance, analysis, and publication of research experiments; and conduct computational and hybrid simulations that may combine the results of multiple distributed experiments and link physical experiments with computer simulations to enable the investigation of overall system performance.

These resources jointly provide the means for collaboration and discovery to improve the seismic design and performance of civil and mechanical infrastructure systems.

Earthquake Simulation

The very first earthquake simulations were performed by statically applying some *horizontal inertia forces* based on scaled peak ground accelerations to a mathematical model of a building. With the further development of computational technologies, static approaches began to give way to dynamic ones.

Dynamic experiments on building and non-building structures may be physical, like shake-table testing, or virtual ones. In both cases, to verify a structure's expected seismic performance, some researchers prefer to deal with so called "real time-histories" though the last cannot be "real" for a hypothetical earthquake specified by either a building code or by some particular research requirements. Therefore, there is a strong incentive to engage an earthquake simulation which is the seismic input that possesses only essential features of a real event.

Sometimes earthquake simulation is understood as a re-creation of local effects of a strong earth shaking.

Structure Simulation

Concurrent experiments with two building models which are *kinematically equivalent* to a real prototype.

Theoretical or experimental evaluation of anticipated seismic performance mostly requires a structure simulation which is based on the concept of structural likeness or similarity. Similarity is some degree of analogy or resemblance between two or more objects. The notion of similarity rests either on exact or approximate repetitions of patterns in the compared items.

In general, a building model is said to have similarity with the real object if the two share *geometric similarity*, *kinematic similarity* and *dynamic similarity*. The most vivid and effective type of similarity is the *kinematic* one. *Kinematic similarity* exists when the paths and velocities of moving particles of a model and its prototype are similar.

The ultimate level of *kinematic similarity* is *kinematic equivalence* when, in the case of earthquake engineering, time-histories of each story lateral displacements of the model and its prototype would be the same.

Seismic Vibration Control

Seismic vibration control is a set of technical means aimed to mitigate seismic impacts in building and non-building structures. All seismic vibration control devices may be classified as *passive*, *active* or *hybrid* where:

- *passive control devices* have no feedback capability between them, structural elements and the ground;
- *active control devices* incorporate real-time recording instrumentation on the ground integrated with earthquake input processing equipment and actuators within the structure;
- *hybrid control devices* have combined features of active and passive control systems.

When ground seismic waves reach up and start to penetrate a base of a building, their energy flow density, due to reflections, reduces dramatically: usually, up to 90%. However, the remaining portions of the incident waves during a major earthquake still bear a huge devastating potential.

After the seismic waves enter a superstructure, there are a number of ways to control them in order to soothe their damaging effect and improve the building's seismic performance, for instance:

- to dissipate the wave energy inside a superstructure with properly engineered dampers;
- to disperse the wave energy between a wider range of frequencies;
- to absorb the resonant portions of the whole wave frequencies band with the help of so-called *mass dampers*.

Mausoleum of Cyrus, the oldest base-isolated structure in the world

Devices of the last kind, abbreviated correspondingly as TMD for the tuned (*passive*), as AMD for the *active*, and as HMD for the *hybrid mass dampers*, have been studied and installed in high-rise buildings, predominantly in Japan, for a quarter of a century.

However, there is quite another approach: partial suppression of the seismic energy flow into the superstructure known as seismic or base isolation.

For this, some pads are inserted into or under all major load-carrying elements in the base of the building which should substantially decouple a superstructure from its substructure resting on a shaking ground.

The first evidence of earthquake protection by using the principle of base isolation was discovered in Pasargadae, a city in ancient Persia, now Iran, and dates back to the 6th century BCE. Below, there are some samples of seismic vibration control technologies of today.

Dry-Stone Walls Control

Dry-stone walls of Machu Picchu Temple of the Sun, Peru

People of Inca civilization were masters of the polished 'dry-stone walls', called ashlar, where blocks of stone were cut to fit together tightly without any mortar. The Incas were among the best stonemasons the world has ever seen and many junctions in their masonry were so perfect that even blades of grass could not fit between the stones.

Peru is a highly seismic land and for centuries the mortar-free construction proved to be apparently more earthquake-resistant than using mortar. The stones of the dry-stone walls built by the Incas could move slightly and resettle without the walls collapsing, a passive structural control technique employing both the principle of energy dissipation and that of suppressing resonant amplifications.

Lead Rubber Bearing

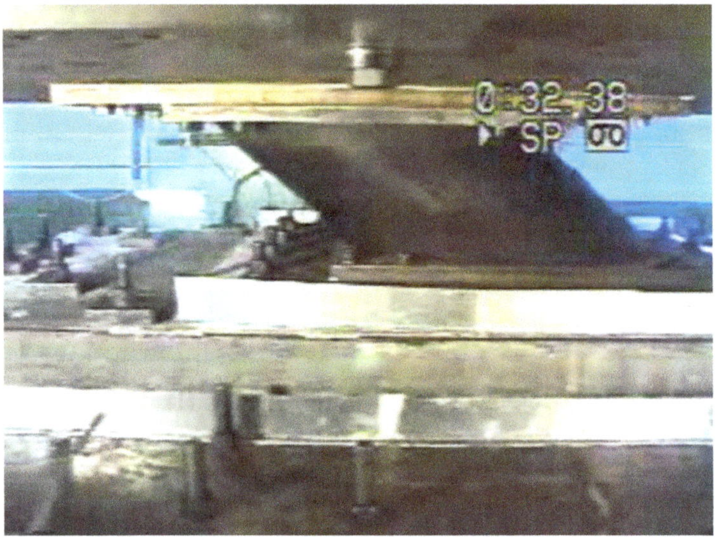

LRB being tested at the UCSD Caltrans-SRMD facility

Lead Rubber Bearing or LRB is a type of base isolation employing a heavy damping. It was invented by Bill Robinson, a New Zealander.

Heavy damping mechanism incorporated in vibration control technologies and, particularly, in base isolation devices, is often considered a valuable source of suppressing vibrations thus enhancing a building's seismic performance. However, for the rather pliant systems such as base isolated structures, with a relatively low bearing stiffness but with a high damping, the so-called "damping force" may turn out the main pushing force at a strong earthquake. The video shows a Lead Rubber Bearing being tested at the UCSD Caltrans-SRMD facility. The bearing is made of rubber with a lead core. It was a uniaxial test in which the bearing was also under a full structure load. Many buildings and bridges, both in New Zealand and elsewhere, are protected with lead dampers and lead and rubber bearings. Te Papa Tongarewa, the national museum of New Zealand, and the New Zealand Parliament Buildings have been fitted with the bearings. Both are in Wellington which sits on an active earthquake fault.

Tuned Mass Damper

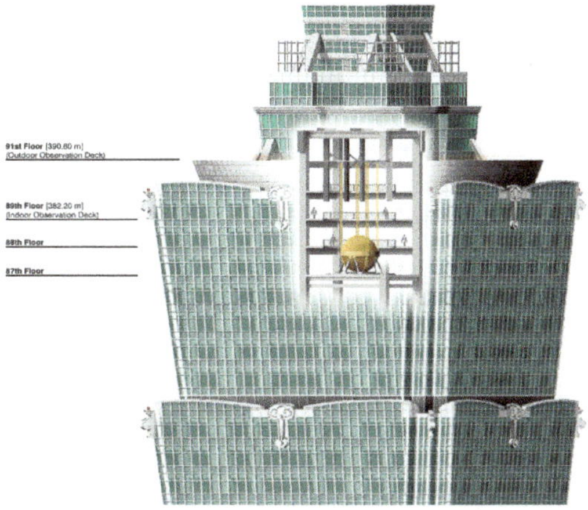

Tuned mass damper in Taipei 101, the world's third tallest skyscraper

Typically the tuned mass dampers are huge concrete blocks mounted in skyscrapers or other structures and moved in opposition to the resonance frequency oscillations of the structures by means of some sort of spring mechanism.

Taipei 101 skyscraper needs to withstand typhoon winds and earthquake tremors common in its area of the Asia-Pacific. For this purpose, a steel pendulum weighing 660 metric tones that serves as a tuned mass damper was designed and installed atop the structure. Suspended from the 92nd to the 88th floor, the pendulum sways to decrease resonant amplifications of lateral displacements in the building caused by earthquakes and strong gusts.

Friction Pendulum Bearing

FPB shake-table testing

Friction Pendulum Bearing (FPB) is another name of Friction Pendulum System (FPS). It is based on three pillars:

- articulated friction slider;
- spherical concave sliding surface;
- enclosing cylinder for lateral displacement restraint.

Snapshot with the link to video clip of a shake-table testing of FPB system supporting a rigid building model is presented at the right.

Building Elevation Control

Transamerica Pyramid building

Building elevation control is a valuable source of vibration control of seismic loading. Pyramid-shaped skyscrapers continue to attract the attention of architects and engineers because such structures promise a better stability against earthquakes and winds. The elevation configuration can prevent buildings' resonant amplifications because a properly configured building disperses the shear wave energy between a wide range of frequencies.

Earthquake or wind quieting ability of the elevation configuration is provided by a specific pattern of multiple reflections and transmissions of vertically propagating waves, which are generated by breakdowns into homogeneity of story layers, and a taper. Any abrupt changes of the propagating waves velocity result in a considerable dispersion of the wave energy between a wide ranges of frequencies thus preventing the resonant displacement amplifications in the building.

A tapered profile of a building is not a compulsory feature of this method of structural control. A similar resonance preventing effect can be also obtained by a proper *tapering* of other characteristics of a building structure, namely, its mass and stiffness. As a result, the building elevation configuration techniques permit an architectural design that may be both attractive and functional.

Simple Roller Bearing

Simple roller bearing is a base isolation device which is intended for protection of various building and non-building structures against potentially damaging lateral impacts of strong earthquakes.

This metallic bearing support may be adapted, with certain precautions, as a seismic isolator to skyscrapers and buildings on soft ground. Recently, it has been employed under the name of *Metallic Roller Bearing* for a housing complex (17 stories) in Tokyo, Japan.

Springs-with-Damper Base Isolator

Springs-with-damper close-up

Springs-with-damper base isolator installed under a three-story town-house, Santa Monica, California is shown on the photo taken prior to the 1994 Northridge earthquake exposure. It is a base isolation device conceptually similar to *Lead Rubber Bearing*.

One of two three-story town-houses like this, which was well instrumented for recording of both vertical and horizontal accelerations on its floors and the ground, has survived a severe shaking during the Northridge earthquake and left valuable recorded information for further study.

Hysteretic Damper

Hysteretic damper is intended to provide better and more reliable seismic performance than that of a conventional structure at the expense of the seismic input energy dissipation. There are five major groups of hysteretic dampers used for the purpose, namely:

- Fluid viscous dampers (FVDs)
- Metallic yielding dampers (MYDs)
- Viscoelastic dampers (VEDs)
- Friction dampers (FDs)
- Straddlingpendulum dampers (swing)

Each group of dampers has specific characteristics, advantages and disadvantages for structural applications.

Seismic Design

Seismic design is based on authorized engineering procedures, principles and criteria meant to design or retrofit structures subject to earthquake exposure. Those criteria are only consistent with the contemporary state of the knowledge about earthquake engineering structures. Therefore, a building design which exactly follows seismic code regulations does not guarantee safety against collapse or serious damage.

The price of poor seismic design may be enormous. Nevertheless, seismic design has always been a trial and error process whether it was based on physical laws or on empirical knowledge of the structural performance of different shapes and materials.

San Francisco City Hall destroyed by 1906 earthquake and fire.

To practice seismic design, seismic analysis or seismic evaluation of new and existing civil engineering projects, an engineer should, normally, pass examination on *Seismic Principles* which, in the State of California, include:

- Seismic Data and Seismic Design Criteria
- Seismic Characteristics of Engineered Systems
- Seismic Forces
- Seismic Analysis Procedures
- Seismic Detailing and Construction Quality Control

San Francisco after the 1906 earthquake and fire

To build up complex structural systems, seismic design largely uses the same relatively small number of basic structural elements (to say nothing of vibration control devices) as any non-seismic design project.

Normally, according to building codes, structures are designed to "withstand" the largest earthquake of a certain probability that is likely to occur at their location. This means the loss of life should be minimized by preventing collapse of the buildings.

Seismic design is carried out by understanding the possible failure modes of a structure and providing the structure with appropriate strength, stiffness, ductility, and configuration to ensure those modes cannot occur.

Seismic Design Requirements

Seismic design requirements depend on the type of the structure, locality of the project and its authorities which stipulate applicable seismic design codes and criteria. For instance, California Department of Transportation's requirements called *The Seismic Design Criteria* (SDC) and aimed at the design of new bridges in California incorporate an innovative seismic performance-based approach.

The Metsamor Nuclear Power Plant was closed after the 1988 Armenian earthquake

The most significant feature in the SDC design philosophy is a shift from a *force-based assessment* of seismic demand to a *displacement-based assessment* of demand and capacity. Thus, the newly adopted displacement approach is based on comparing the *elastic displacement* demand to the *inelastic displacement* capacity of the primary structural components while ensuring a minimum level of inelastic capacity at all potential plastic hinge locations.

In addition to the designed structure itself, seismic design requirements may include a *ground stabilization* underneath the structure: sometimes, heavily shaken ground breaks up which leads to collapse of the structure sitting upon it. The following topics should be of primary concerns: liquefaction; dynamic lateral earth pressures on retaining walls; seismic slope stability; earthquake-induced settlement.

Nuclear facilities should not jeopardise their safety in case of earthquakes or other hostile external events. Therefore, their seismic design is based on criteria far more stringent than those applying to non-nuclear facilities. The Fukushima I nuclear accidents and damage to other nuclear facilities that followed the 2011 Tōhoku earthquake and tsunami have, however, drawn attention to ongoing concerns over Japanese nuclear seismic design standards and caused other many governments to re-evaluate their nuclear programs. Doubt has also been expressed over the seismic evaluation and design of certain other plants, including the Fessenheim Nuclear Power Plant in France.

Failure Modes

Failure mode is the manner by which an earthquake induced failure is observed. It, generally, describes the way the failure occurs. Though costly and time consuming, learning from each real earthquake failure remains a routine recipe for advancement in *seismic design* methods. Below, some typical modes of earthquake-generated failures are presented.

Typical damage to unreinforced masonry buildings at earthquakes

The lack of reinforcement coupled with poor mortar and inadequate roof-to-wall ties can result in substantial damage to an unreinforced masonry building. Severely cracked or leaning walls are some of the most common earthquake damage. Also hazardous is the damage that may occur between the walls and roof or floor diaphragms. Separation between the framing and the walls can jeopardize the vertical support of roof and floor systems.

Soft story collapse due to inadequate shear strength at ground level, Loma Prieta earthquake

Soft story effect. Absence of adequate shear walls on the ground level caused damage to this structure. A close examination of the image reveals that the rough board siding, once covered by a brick veneer, has been completely dismantled from the studwall. Only the rigidity of the floor above combined with the support on the two hidden sides by continuous walls, not penetrated with large doors as on the street sides, is preventing full collapse of the structure.

Effects of soil liquefaction during the 1964 Niigata earthquake

Soil liquefaction. In the cases where the soil consists of loose granular deposited materials with the tendency to develop excessive hydrostatic pore water pressure of sufficient magnitude and compact, liquefaction of those loose saturated deposits may result in non-uniform settlements and tilting of structures. This caused major damage to thousands of buildings in Niigata, Japan during the 1964 earthquake.

Car smashed by landslide rock, 2008 Sichuan earthquake

Landslide rock fall. A landslide is a geological phenomenon which includes a wide range of ground movement, including rock falls. Typically, the action of gravity is the primary driving force for a landslide to occur though in this case there was another contributing factor which affected the original slope stability: the landslide required an *earthquake trigger* before being released.

Effects of pounding against adjacent building, Loma Prieta

Pounding against adjacent building. This is a photograph of the collapsed five-story tower, St. Joseph's Seminary, Los Altos, California which resulted in one fatality. During Loma Prieta earthquake, the tower pounded against the independently vibrating adjacent building behind. A possibility of pounding depends on both buildings' lateral displacements which should be accurately estimated and accounted for.

Effects of completely shattered joints of concrete frame, Northridge

At Northridge earthquake, the Kaiser Permanente concrete frame office building had joints completely shattered, revealing inadequate confinement steel, which resulted in the second story collapse. In the transverse direction, composite end shear walls, consisting of two wythes of brick and a layer of shotcrete that carried the lateral load, peeled apart because of inadequate through-ties and failed.

7-story reinforced concrete buildings on steep slope collapse due to the following:

- Improper construction site on a foothill.
- Poor detailing of the reinforcement (lack of concrete confinement in the columns and at the beam-column joints, inadequate splice length).
- Seismically weak soft story at the first floor.
- Long cantilevers with heavy dead load.

shifting from foundation, Whittier

Sliding off foundations effect of a relatively rigid residential building structure during 1987 Whittier Narrows earthquake. The magnitude 5.9 earthquake pounded the Garvey West Apartment building in Monterey Park, California and shifted its superstructure about 10 inches to the east on its foundation.

Earthquake damage in Pichilemu.

If a superstructure is not mounted on a base isolation system, its shifting on the basement should be prevented.

Insufficient shear reinforcement let main rebars to buckle, Northridge

Reinforced concrete column burst at Northridge earthquake due to insufficient shear reinforcement mode which allows main reinforcement to buckle outwards. The deck unseated at the hinge and failed in shear. As a result, the La Cienega-Venice underpass section of the 10 Freeway collapsed.

Support-columns and upper deck failure, Loma Prieta earthquake

Loma Prieta earthquake: side view of reinforced concrete support-columns failure which triggered the upper deck collapse onto the lower deck of the two-level Cypress viaduct of Interstate Highway 880, Oakland, CA.

Failure of retaining wall due to ground movement, Loma Prieta

Retaining wall failure at Loma Prieta earthquake in Santa Cruz Mountains area: prominent northwest-trending extensional cracks up to 12 cm (4.7 in) wide in the concrete spillway to Austrian Dam, the north abutment.

Lateral spreading mode of ground failure, Loma Prieta

Ground shaking triggered soil liquefaction in a subsurface layer of sand, producing differential lateral and vertical movement in an overlying carapace of unliquified sand and silt. This mode of ground failure, termed lateral spreading, is a principal cause of liquefaction-related earthquake damage.

Beams and pier columns diagonal cracking, 2008 Sichuan earthquake

Severely damaged building of Agriculture Development Bank of China after 2008 Sichuan earthquake: most of the beams and pier columns are sheared. Large diagonal cracks in masonry and veneer are due to in-plane loads while abrupt settlement of the right end of the building should be attributed to a landfill which may be hazardous even without any earthquake.

Tsunami strikes Ao Nang,

Twofold tsunami impact: sea waves hydraulic pressure and inundation. Thus, the Indian Ocean earthquake of December 26, 2004, with the epicenter off the west coast of Sumatra, Indonesia, triggered a series of devastating tsunamis, killing more than 230,000 people in eleven countries by inundating surrounding coastal communities with huge waves up to 30 meters (100 feet) high.

Earthquake-Resistant Construction

Earthquake construction means implementation of seismic design to enable building and non-building structures to live through the anticipated earthquake exposure up to the expectations and in compliance with the applicable building codes.

Construction of Pearl River Tower X-bracing to resist lateral forces of earthquakes and winds

Design and construction are intimately related. To achieve a good workmanship, detailing of the members and their connections should be as simple as possible. As any construction in general, earthquake construction is a process that consists of the building, retrofitting or assembling of infrastructure given the construction materials available.

The destabilizing action of an earthquake on constructions may be *direct* (seismic motion of the ground) or *indirect* (earthquake-induced landslides, soil liquefaction and waves of tsunami).

A structure might have all the appearances of stability, yet offer nothing but danger when an earthquake occurs. The crucial fact is that, for safety, earthquake-resistant construction techniques are as important as quality control and using correct materials. *Earthquake contractor* should be registered in the state of the project location, bonded and insured.

To minimize possible losses, construction process should be organized with keeping in mind that earthquake may strike any time prior to the end of construction.

Each construction project requires a qualified team of professionals who understand the basic features of seismic performance of different structures as well as construction management.

Adobe Structures

Partially collapsed adobe building in Westmorland, California

Around thirty percent of the world's population lives or works in earth-made construction. Adobe type of mud bricks is one of the oldest and most widely used building materials. The use of adobe is very common in some of the world's most hazard-prone regions, traditionally across Latin America, Africa, Indian subcontinent and other parts of Asia, Middle East and Southern Europe.

Adobe buildings are considered very vulnerable at strong quakes. However, multiple ways of seismic strengthening of new and existing adobe buildings are available.

Key factors for the improved seismic performance of adobe construction are:

- Quality of construction.
- Compact, box-type layout.
- Seismic reinforcement.

Limestone and Sandstone Structures

Base-isolated City and County Building, Salt Lake City, Utah

Limestone is very common in architecture, especially in North America and Europe. Many landmarks across the world are made of limestone. Many medieval churches and castles in Europe are made of limestone and sandstone masonry. They are the long-lasting materials but their rather heavy weight is not beneficial for adequate seismic performance.

Application of modern technology to seismic retrofitting can enhance the survivability of unreinforced masonry structures. As an example, from 1973 to 1989, the Salt Lake City and County Building in Utah was exhaustively renovated and repaired with an emphasis on preserving historical accuracy in appearance. This was done in concert with a seismic upgrade that placed the weak sandstone structure on base isolation foundation to better protect it from earthquake damage.

Timber Frame Structures

Anne Hvide's House, Denmark (1560)

Timber framing dates back thousands of years, and has been used in many parts of the world during various periods such as ancient Japan, Europe and medieval England in localities where timber was in good supply and building stone and the skills to work it were not.

The use of timber framing in buildings provides their complete skeletal framing which offers some structural benefits as the timber frame, if properly engineered, lends itself to better *seismic survivability*.

Light-Frame Structures

A two-story wooden-frame for a residential building structure

Light-frame structures usually gain seismic resistance from rigid plywood shear walls and wood structural panel diaphragms. Special provisions for seismic load-resisting systems for all engineered wood structures requires consideration of diaphragm ratios, horizontal and vertical diaphragm shears, and connector/fastener values. In addition, collectors, or drag struts, to distribute shear along a diaphragm length are required.

Reinforced Masonry Structures

A construction system where steel reinforcement is embedded in the mortar joints of masonry or placed in holes and after filled with concrete or grout is called reinforced masonry.

Reinforced hollow masonry wall

The devastating 1933 Long Beach earthquake revealed that masonry construction should be improved immediately. Then, the California State Code made the reinforced masonry mandatory.

There are various practices and techniques to achieve reinforced masonry. The most common type is the reinforced hollow unit masonry. The effectiveness of both vertical and horizontal reinforcement strongly depends on the type and quality of the masonry, i.e. masonry units and mortar.

To achieve a ductile behavior of masonry, it is necessary that the shear strength of the wall is greater than the flexural strength.

Reinforced Concrete Structures

Stressed Ribbon pedestrian bridge over the Rogue River, Grants Pass, Oregon

Reinforced concrete is concrete in which steel reinforcement bars (rebars) or fibers have been incorporated to strengthen a material that would otherwise be brittle. It can be used to produce beams, columns, floors or bridges.

Prestressed concrete cable-stayed bridge over Yangtze river

Prestressed concrete is a kind of reinforced concrete used for overcoming concrete's natural weakness in tension. It can be applied to beams, floors or bridges with a longer span than is practical with ordinary reinforced concrete. Prestressing tendons (generally of high tensile steel cable or rods) are used to provide a clamping load which produces a compressive stress that offsets the tensile stress that the concrete compression member would, otherwise, experience due to a bending load.

To prevent catastrophic collapse in response earth shaking (in the interest of life safety), a traditional reinforced concrete frame should have ductile joints. Depending upon the methods used and the imposed seismic forces, such buildings may be immediately usable, require extensive repair, or may have to be demolished.

Prestressed Structures

Prestressed structure is the one whose overall integrity, stability and security depend, primarily, on a *prestressing*. *Prestressing* means the intentional creation of permanent stresses in a structure for the purpose of improving its performance under various service conditions.

Naturally pre-compressed exterior wall of Colosseum, Rome

There are the following basic types of prestressing:

- Pre-compression (mostly, with the own weight of a structure)
- Pretensioning with high-strength embedded tendons
- Post-tensioning with high-strength bonded or unbonded tendons

Today, the concept of prestressed structure is widely engaged in design of buildings, underground structures, TV towers, power stations, floating storage and offshore facilities, nuclear reactor vessels, and numerous kinds of bridge systems.

A beneficial idea of *prestressing* was, apparently, familiar to the ancient Rome architects; look, e.g., at the tall attic wall of Colosseum working as a stabilizing device for the wall piers beneath.

Steel Structures

Collapsed section of the San Francisco–Oakland Bay Bridge in response to Loma Prieta earthquake

Steel structures are considered mostly earthquake resistant but some failures have occurred. A great number of welded steel moment-resisting frame buildings, which looked earthquake-proof, surprisingly experienced brittle behavior and were hazardously damaged in the 1994 Northridge earthquake. After that, the Federal Emergency Management Agency (FEMA) initiated development of repair techniques and new design approaches to minimize damage to steel moment frame buildings in future earthquakes.

For structural steel seismic design based on Load and Resistance Factor Design (LRFD) approach, it is very important to assess ability of a structure to develop and maintain its bearing resistance in the inelastic range. A measure of this ability is ductility, which may be observed in a *material itself*, in a *structural element*, or to a *whole structure*.

As a consequence of Northridge earthquake experience, the American Institute of Steel Construction has introduced AISC 358 "Pre-Qualified Connections for Special and intermediate Steel Moment Frames." The AISC Seismic Design Provisions require that all Steel Moment Resisting Frames employ either connections contained in AISC 358, or the use of connections that have been subjected to pre-qualifying cyclic testing.

Prediction of Earthquake Losses

Earthquake loss estimation is usually defined as a *Damage Ratio* (DR) which is a ratio of the earthquake damage repair cost to the total value of a building. *Probable Maximum Loss* (PML) is a common term used for earthquake loss estimation, but it lacks a precise definition. In 1999, ASTM E2026 'Standard Guide for the Estimation of Building Damageability in Earthquakes' was produced in order to standardize the nomenclature for seismic loss estimation, as well as establish guidelines as to the review process and qualifications of the reviewer.

Earthquake loss estimations are also referred to as *Seismic Risk Assessments*. The risk assessment process generally involves determining the probability of various ground motions coupled with the vulnerability or damage of the building under those ground motions. The results are defined as a percent of building replacement value.

Structural Engineering

Structural engineering deals with the making of complex systems like the International Space Station, seen here from the departing Space Shuttle *Atlantis*.

Structural engineers investigating NASA's Mars-bound spacecraft, the Phoenix Mars Lander

Structural engineers are trained to understand and calculate the stability, strength and rigidity of built structures for buildings and nonbuilding structures, to develop designs and integrate their design with that of other designers, and to supervise construction of projects on site. They can also be involved in the design of machinery, medical equipment, vehicles etc. where structural integrity affects functioning and safety.

The Eiffel Tower is a historical achievement of structural engineering.

Structural engineering theory is based upon applied physical laws and empirical knowledge of the structural performance of different materials and geometries. Structural engineering design utilizes a number of relatively simple structural elements to build complex structural systems. Structural engineers are responsible for making creative and efficient use of funds, structural elements and materials to achieve these goals.

Structural Engineer (Professional)

Structural engineers are responsible for engineering design and analysis. Entry-level structural engineers may design the individual structural elements of a structure, for example the beams, columns, and floors of a building. More experienced engineers may be responsible for the structural design and integrity of an entire system, such as a building.

Structural engineers often specialize in particular fields, such as bridge engineering, building engineering, pipeline engineering, industrial structures, or special mechanical structures such as vehicles, ships or aircraft.

Structural engineering has existed since humans first started to construct their own structures. It became a more defined and formalised profession with the emergence of the architecture profession as distinct from the engineering profession during the industrial revolution in the late 19th century. Until then, the architect and the structural engineer were usually one and the same - the master builder. Only with the development of specialised knowledge of structural theories that emerged during the 19th and early 20th centuries did the professional structural engineer come into existence.

The role of a structural engineer today involves a significant understanding of both static and dynamic loading, and the structures that are available to resist them. The complexity of modern structures often requires a great deal of creativity from the engineer in order to ensure the structures support and resist the loads they are subjected to. A structural engineer will typically have a four or five year undergraduate degree, followed by a minimum of three years of professional practice before being considered fully qualified. Structural engineers are licensed or accredited by different learned societies and regulatory bodies around the world (for example, the Institution of Structural Engineers in the UK). Depending on the degree course they have studied and/or the jurisdiction they are seeking licensure in, they may be accredited (or licensed) as just structural engineers, or as civil engineers, or as both civil and structural engineers. Another international organisation is IABSE (International Association for Bridge and Structural Engineering). The aim of that association is to exchange knowledge and to advance the practice of structural engineering worldwide in the service of the profession and society.

History of Structural Engineering

Pont du Gard, France, a Roman era aqueduct circa 19 BC.

Structural engineering dates back to 2700 B.C.E. when the step pyramid for Pharaoh Djoser was built by Imhotep, the first engineer in history known by name. Pyramids were the most common major structures built by ancient civilizations because the structural form of a pyramid is inherently stable and can be almost infinitely scaled (as opposed to most other structural forms, which cannot be linearly increased in size in proportion to increased loads).

However, it's important to note that the structural stability of the pyramid is not primarily a result of its shape. The integrity of the pyramid is intact as long as each of the stones is able to support the weight of the stone above it. The limestone blocks were taken from a quarry near the build site. Since the compressive strength of limestone is anywhere from 30 to 250 MPa (MPa = Pa * 10^6), the blocks will not fail under compression. Therefore, the structural strength of the pyramid stems from the material properties of the stones from which it was built rather than the pyramid's geometry.

Throughout ancient and medieval history most architectural design and construction was carried out by artisans, such as stone masons and carpenters, rising to the role of master builder. No theory of structures existed, and understanding of how structures stood up was extremely limited, and based almost entirely on empirical evidence of 'what had worked before'. Knowledge was retained by guilds and seldom supplanted by advances. Structures were repetitive, and increases in scale were incremental.

No record exists of the first calculations of the strength of structural members or the behavior of structural material, but the profession of structural engineer only really took shape with the Industrial Revolution and the re-invention of concrete. The physical sciences underlying structural engineering began to be understood in the Renaissance and have since developed into computer-based applications pioneered in the 1970s.

Timeline

- 1452–1519 Leonardo da Vinci made many contributions
- 1638: Galileo Galilei published the book "Two New Sciences" in which he examined the failure of simple structures

Galileo Galilei published the book "Two New Sciences" in which he examined the failure of simple structures

- 1660: Hooke's law by Robert Hooke
- 1687: Isaac Newton published "Philosophiae Naturalis Principia Mathematica" which contains the Newton's laws of motion

Isaac Newton published "Philosophiae Naturalis Principia Mathematica" which contains the Newton's laws of motion

- 1750: Euler–Bernoulli beam equation
- 1700–1782: Daniel Bernoulli introduced the principle of virtual work
- 1707–1783: Leonhard Euler developed the theory of buckling of columns

Leonhard Euler developed the theory of buckling of columns

- 1826: Claude-Louis Navier published a treatise on the elastic behaviors of structures
- 1873: Carlo Alberto Castigliano presented his dissertation "Intorno ai sistemi elastici", which contains his theorem for computing displacement as partial derivative of the strain energy. This theorem includes the method of *least work* as a special case
- 1874: Otto Mohr formalized the idea of a statically indeterminate structure.
- 1922: Timoshenko corrects the Euler-Bernoulli beam equation
- 1936: Hardy Cross' publication of the moment distribution method, an important innovation in the design of continuous frames.
- 1941: Alexander Hrennikoff solved the discretization of plane elasticity problems using a lattice framework
- 1942: R. Courant divided a domain into finite subregions
- 1956: J. Turner, R. W. Clough, H. C. Martin, and L. J. Topp's paper on the "Stiffness and Deflection of Complex Structures" introduces the name "finite-element method" and is widely recognized as the first comprehensive treatment of the method as it is known today

Structural Failure

The history of structural engineering contains many collapses and failures. Sometimes this is due to obvious negligence, as in the case of the Pétionville school collapse, in which Rev. Fortin Augustin *"constructed the building all by himself, saying he didn't need an engineer as he had good knowledge of construction"* following a partial collapse of the three-story schoolhouse that sent neighbors fleeing. The final collapse killed 94 people, mostly children.

In other cases structural failures require careful study, and the results of these inquiries have resulted in improved practices and greater understanding of the science of structural engineering. Some such studies are the result of forensic engineering investigations where the original engineer seems to have done everything in accordance with the state of the profession and acceptable practice yet a failure still eventuated. A famous case of structural knowledge and practice being advanced in this manner can be found in a series of failures involving box girders which collapsed in Australia during the 1970s.

Specializations

Building Structures

Sydney Opera House, designed by Architect Jørn Utzon and structural design by Ove Arup & Partners

Millennium Dome in London, UK, by Richard Rogers and Buro Happold

Burj Khalifa, in Dubai, the world's tallest building, shown under construction in 2007 (since completed)

Structural building engineering includes all structural engineering related to the design of buildings. It is a branch of structural engineering closely affiliated with architecture.

Structural building engineering is primarily driven by the creative manipulation of materials and forms and the underlying mathematical and scientific ideas to achieve an end which fulfills its functional requirements and is structurally safe when subjected to all the loads it could reasonably be expected to experience. This is subtly different from architectural design, which is driven by the creative manipulation of materials and forms, mass, space, volume, texture and light to achieve an end which is aesthetic, functional and often artistic.

The architect is usually the lead designer on buildings, with a structural engineer employed as a sub-consultant. The degree to which each discipline actually leads the design depends heavily on the type of structure. Many structures are structurally simple and led by architecture, such as multi-storey office buildings and housing, while other structures, such as tensile structures, shells and gridshells are heavily dependent on their form for their strength, and the engineer may have a more significant influence on the form, and hence much of the aesthetic, than the architect.

The structural design for a building must ensure that the building is able to stand up safely, able to function without excessive deflections or movements which may cause fatigue of structural elements, cracking or failure of fixtures, fittings or partitions, or discomfort for occupants. It must account for movements and forces due to temperature, creep, cracking and imposed loads. It must also ensure that the design is practically buildable within acceptable manufacturing tolerances of the materials. It must allow the architecture to work, and the building services to fit within the building and function (air conditioning, ventilation, smoke extract, electrics, lighting etc.). The structural design of a modern building can be extremely complex, and often requires a large team to complete.

Structural engineering specialties for buildings include:

- Earthquake engineering
- Façade engineering
- Fire engineering
- Roof engineering
- Tower engineering
- Wind engineering

Earthquake Engineering Structures

Earthquake engineering structures are those engineered to withstand earthquakes.

Earthquake-proof pyramid El Castillo, Chichen Itza

The main objectives of earthquake engineering are to understand the interaction of structures with the shaking ground, foresee the consequences of possible earthquakes, and design and construct the structures to perform during an earthquake.

Earthquake-proof structures are not necessarily extremely strong like the El Castillo pyramid at Chichen Itza shown above. In fact, many structures considered strong may in fact be stiff, which can result in poor seismic performance.

One important tool of earthquake engineering is base isolation, which allows the base of a structure to move freely with the ground.

Civil Engineering Structures

Civil structural engineering includes all structural engineering related to the built environment. It includes:

		• Power stations
• Bridges		• Railways
• Dams		• Retaining structures and walls
• Earthworks		• Roads
• Foundations		• Tunnels
• Offshore structures		• Waterways
• Pipelines		• reservoir
		• Water and wastewater infrastructure

The structural engineer is the lead designer on these structures, and often the sole designer. In the design of structures such as these, structural safety is of paramount importance (in the UK, designs for dams, nuclear power stations and bridges must be signed off by a chartered engineer).

Civil engineering structures are often subjected to very extreme forces, such as large variations in temperature, dynamic loads such as waves or traffic, or high pressures from water or compressed gases. They are also often constructed in corrosive environments, such as at sea, in industrial facilities or below ground.

Mechanical Structures

Mechanical Structures

Principles of structural engineering are applied to variety of mechanical (moveable) structures. The design of static structures assumes they always have the same geometry (in fact, so-called static structures can move significantly, and structural engineering design must take this into account where necessary), but the design of moveable or moving structures must account for fatigue, variation in the method in which load is resisted and significant deflections of structures.

The forces which parts of a machine are subjected to can vary significantly, and can do so at a great rate. The forces which a boat or aircraft are subjected to vary enormously and will do so thousands of times over the structure's lifetime. The structural design must ensure that such structures are able to endure such loading for their entire design life without failing.

These works can require mechanical structural engineering:

- Boilers and pressure vessels
- Coachworks and carriages
- Cranes
- Elevators
- Escalators
- Marine vessels and hulls

Aerospace Structures

An Airbus A380, the world's largest passenger airliner

Design of missile needs in depth understanding of Structural Analysis

Aerospace structure types include launch vehicles, (Atlas, Delta, Titan), missiles (ALCM, Harpoon), Hypersonic vehicles (Space Shuttle), military aircraft (F-16, F-18) and commercial aircraft (Boeing 777, MD-11). Aerospace structures typically consist of thin plates with stiffeners for the external surfaces, bulkheads and frames to support the shape and fasteners such as welds, rivets, screws and bolts to hold the components together.

Nanoscale Structures

A nanostructure is an object of intermediate size between molecular and microscopic (micrometer-sized) structures. In describing nanostructures it is necessary to differentiate between the number of dimensions on the nanoscale. Nanotextured surfaces have one dimension on the nanoscale, i.e., only the thickness of the surface of an object is between 0.1 and 100 nm. Nanotubes have two dimensions on the nanoscale, i.e., the diameter of the tube is between 0.1 and 100 nm; its length could be much greater. Finally, spherical nanoparticles have three dimensions on the nanoscale, i.e., the particle is between 0.1 and 100 nm in each spatial dimension. The terms nanoparticles and ultrafine particles (UFP) often are used synonymously although UFP can reach into the micrometre range. The term 'nanostructure' is often used when referring to magnetic technology.

Structural Engineering for Medical Science

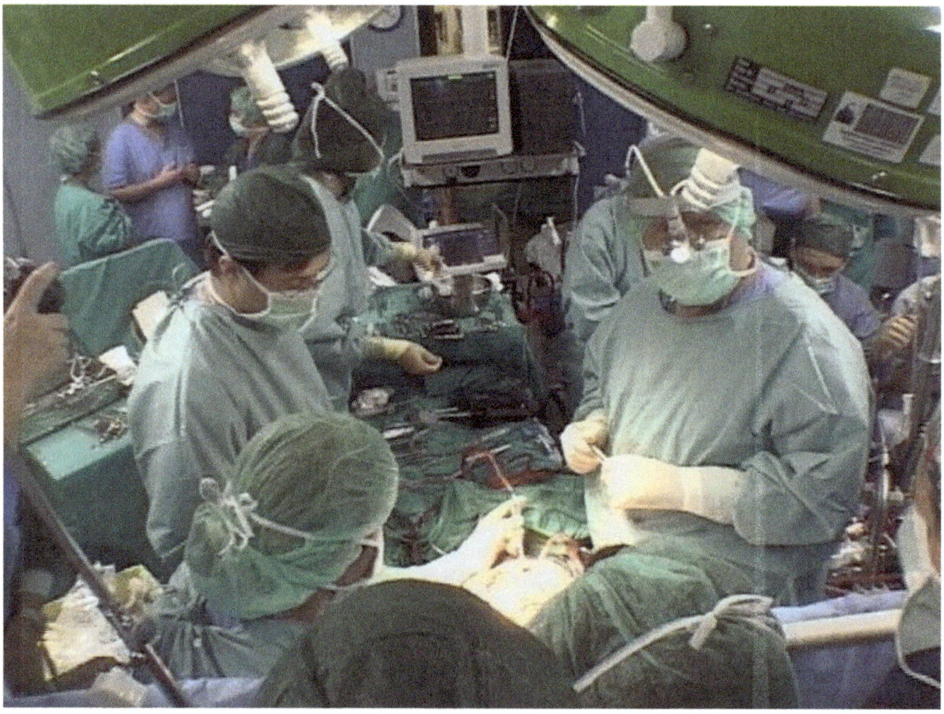

Designing Medical Equipment needs in-depth understanding of Structural Engineering

Medical equipment (also known as armamentarium) is designed to aid in the diagnosis, monitoring or treatment of medical conditions. There are several basic types: Diagnostic equipment includes medical imaging machines, used to aid in diagnosis ; equipment includes infusion pumps, medical lasers and LASIK surgical machines ; Medical monitors allow medical staff to measure a patient's medical state. Monitors may measure patient vital signs and other parameters including ECG, EEG, blood pressure, and dissolved gases in the blood ; Diagnostic Medical Equipment may also be used in the home for certain purposes, e.g. for the control of diabetes mellitus. A biomedical equipment technician (BMET) is a vital component of the healthcare delivery system. Employed primarily by hospitals, BMETs are the people responsible for maintaining a facility's medical equipment.

Structural Elements

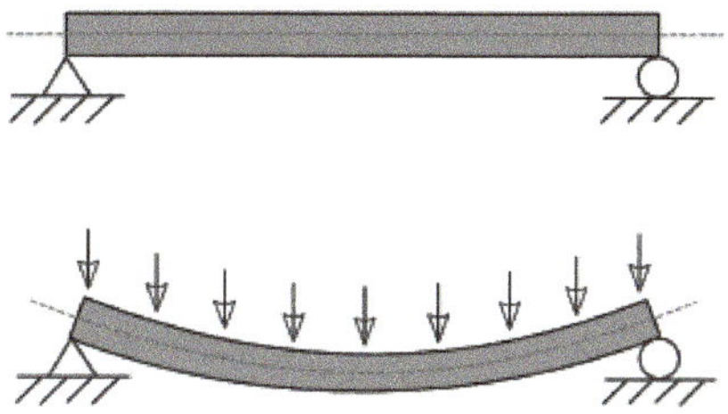

A statically determinate simply supported beam, bending under an evenly distributed load.

Any structure is essentially made up of only a small number of different types of elements:

- Columns
- Beams
- Plates
- Arches
- Shells
- Catenaries

Many of these elements can be classified according to form (straight, plane / curve) and dimensionality (one-dimensional / two-dimensional):

	One-dimensional		Two-dimensional	
	straight	**curve**	**plane**	**curve**
(predominantly) bending	beam	continuous arch	plate, concrete slab	lamina, dome
(predominant) tensile stress	rope, tie	Catenary	shell	
(predominant) compression	pier, column		Load-bearing wall	

Columns

Columns are elements that carry only axial force - compression - or both axial force and bending (which is technically called a beam-column but practically, just a column). The design of a column must check the axial capacity of the element, and the buckling capacity.

The buckling capacity is the capacity of the element to withstand the propensity to buckle. Its capacity depends upon its geometry, material, and the effective length of the column, which depends upon the restraint conditions at the top and bottom of the column. The effective length is $K * l$ where l is the real length of the column and K is the factor dependent on the restraint conditions.

The capacity of a column to carry axial load depends on the degree of bending it is subjected to, and vice versa. This is represented on an interaction chart and is a complex non-linear relationship.

Beams

Little Belt: a truss bridge in Denmark

A beam may be defined as an element in which one dimension is much greater than the other two and the applied loads are usually normal to the main axis of the element. Beams and columns are called line elements and are often represented by simple lines in structural modeling.

- cantilevered (supported at one end only with a fixed connection)
- simply supported (supported vertically at each end; horizontally on only one to withstand friction, and able to rotate at the supports)
- fixed (supported at both ends by fixed connection; unable to rotate at the supports)
- continuous (supported by three or more supports)
- a combination of the above (ex. supported at one end and in the middle)

Beams are elements which carry pure bending only. Bending causes one part of the section of a beam (divided along its length) to go into compression and the other part into tension. The compression part must be designed to resist buckling and crushing, while the tension part must be able to adequately resist the tension.

Trusses

The McDonnell Planetarium by Gyo Obata in St Louis, Missouri, USA, a concrete shell structure

The 630 foot (192 m) high, stainless-clad (type 304) Gateway Arch in Saint Louis, Missouri

A truss is a structure comprising two types of structural elements; compression members and tension members (i.e. struts and ties). Most trusses use gusset plates to connect intersecting elements. Gusset plates are relatively flexible and minimize bending moments at the connections, thus allowing the truss members to carry primarily tension or compression.

Trusses are usually utilised in large-span structures, where it would be uneconomical to use solid beams.

Plates

Plates carry bending in two directions. A concrete flat slab is an example of a plate. Plates are understood by using continuum mechanics, but due to the complexity involved they are most often designed using a codified empirical approach, or computer analysis.

They can also be designed with yield line theory, where an assumed collapse mechanism is analysed to give an upper bound on the collapse load. This technique is used in practice but because the method provides an upper-bound, i.e. an unsafe prediction of the collapse load, for poorly conceived collapse mechanisms great care is needed to ensure that the assumed collapse mechanism is realistic.

Shells

Shells derive their strength from their form, and carry forces in compression in two directions. A dome is an example of a shell. They can be designed by making a hanging-chain model, which will act as a catenary in pure tension, and inverting the form to achieve pure compression.

Arches

Arches carry forces in compression in one direction only, which is why it is appropriate to build arches out of masonry. They are designed by ensuring that the line of thrust of the force remains within the depth of the arch. It is mainly used to increase the bountifulness of any structure.

Catenaries

Catenaries derive their strength from their form, and carry transverse forces in pure tension by deflecting (just as a tightrope will sag when someone walks on it). They are almost always cable or fabric structures. A fabric structure acts as a catenary in two directions.

Structural Engineering Theory

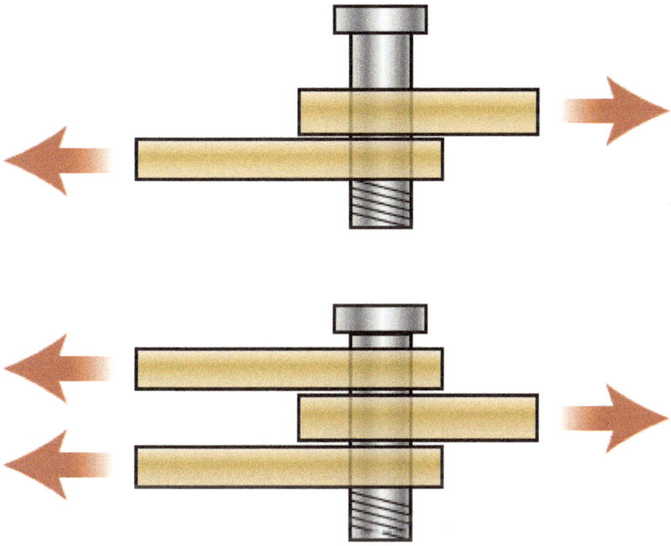

Figure of a bolt in shear stress. Top figure illustrates single shear, bottom figure illustrates double shear.

Structural engineering depends upon a detailed knowledge of applied mechanics, materials science and applied mathematics to understand and predict how structures support and resist self-weight and imposed loads. To apply the knowledge successfully a structural engineer generally requires detailed knowledge of relevant empirical and theoretical design codes, the techniques of structural analysis, as well as some knowledge of the corrosion resistance of the materials and structures, especially when those structures are exposed to the external environment. Since the 1990s, specialist software has become available to aid in the design of structures, with the functionality to assist in the drawing, analyzing and designing of structures with maximum precision; examples include AutoCAD, StaadPro, ETABS, Prokon, Revit Structure etc. Such software may also take into consideration environmental loads, such as from earthquakes and winds.

Materials

Structural engineering depends on the knowledge of materials and their properties, in order to understand how different materials support and resist loads.

Common structural materials are:

- Iron: Wrought iron, Cast iron
- Concrete: Reinforced concrete, Prestressed concrete
- Alloy: Steel, Stainless steel

- Masonry
- Timber: Hardwood, Softwood
- Aluminium
- Composite materials: Plywood
- Other structural materials:Adobe, Bamboo, Carbon fibre, Fiber reinforced plastic, Mudbrick, Roofing materials

Geotechnical Engineering

Boston's Big Dig presented geotechnical challenges in an urban environment.

Geotechnical engineering is the branch of civil engineering concerned with the engineering behavior of earth materials. Geotechnical engineering is important in civil engineering, but also has applications in military, mining, petroleum and other engineering disciplines that are concerned with construction occurring on the surface or within the ground. Geotechnical engineering uses principles of soil mechanics and rock mechanics to investigate subsurface conditions and materials; determine the relevant physical/mechanical and chemical properties of these materials; evaluate stability of natural slopes and man-made soil deposits; assess risks posed by site conditions; design earthworks and structure foundations; and monitor site conditions, earthwork and foundation construction.

A typical geotechnical engineering project begins with a review of project needs to define the required material properties. Then follows a site investigation of soil, rock, fault distribution and bedrock properties on and below an area of interest to determine their engineering properties including how they will interact with, on or in a proposed construction. Site investigations are needed to gain an understanding of the area in or on which the engineering will take place. Investigations can include the assessment of the risk to humans, property and the environment from natural hazards such as earthquakes, landslides, sinkholes, soil liquefaction, debris flows and rockfalls.

A geotechnical engineer then determines and designs the type of foundations, earthworks, and/or pavement subgrades required for the intended man-made structures to be built. Foundations are designed and constructed for structures of various sizes such as high-rise buildings, bridges, medium to large commercial buildings, and smaller structures where the soil conditions do not allow code-based design.

Foundations built for above-ground structures include shallow and deep foundations. Retaining structures include earth-filled dams and retaining walls. Earthworks include embankments, tunnels, dikes and levees, channels, reservoirs, deposition of hazardous waste and sanitary landfills.

Geotechnical engineering is also related to coastal and ocean engineering. Coastal engineering can involve the design and construction of wharves, marinas, and jetties. Ocean engineering can involve foundation and anchor systems for offshore structures such as oil platforms.

The fields of geotechnical engineering and engineering geology are closely related, and have large areas of overlap. However, the field of geotechnical engineering is a specialty of engineering, where the field of engineering geology is a specialty of geology.

History

Humans have historically used soil as a material for flood control, irrigation purposes, burial sites, building foundations, and as construction material for buildings. First activities were linked to irrigation and flood control, as demonstrated by traces of dykes, dams, and canals dating back to at least 2000 BCE that were found in ancient Egypt, ancient Mesopotamia and the Fertile Crescent, as well as around the early settlements of Mohenjo Daro and Harappa in the Indus valley. As the cities expanded, structures were erected supported by formalized foundations; Ancient Greeks notably constructed pad footings and strip-and-raft foundations. Until the 18th century, however, no theoretical basis for soil design had been developed and the discipline was more of an art than a science, relying on past experience.

Several foundation-related engineering problems, such as the Leaning Tower of Pisa, prompted scientists to begin taking a more scientific-based approach to examining the subsurface. The earliest advances occurred in the development of earth pressure theories for the construction of retaining walls. Henri Gautier, a French Royal Engineer, recognized the "natural slope" of different soils in 1717, an idea later known as the soil's angle of repose. A rudimentary soil classification system was also developed based on a material's unit weight, which is no longer considered a good indication of soil type.

The application of the principles of mechanics to soils was documented as early as 1773 when Charles Coulomb (a physicist, engineer, and army Captain) developed improved methods to determine the earth pressures against military ramparts. Coulomb observed that, at failure, a distinct slip plane would form behind a sliding retaining wall and he suggested that the maximum shear stress on the slip plane, for design purposes, was the sum of the soil cohesion, c, and friction $\sigma \tan(\phi)$, where σ is the normal stress on the slip plane and ϕ is the friction angle of the soil. By combining Coulomb's theory with Christian Otto Mohr's 2D stress state, the theory became known as Mohr-Coulomb theory. Although it is now recognized that precise determination of cohesion is impossible because c is not a fundamental soil property, the Mohr-Coulomb theory is still used in practice today.

In the 19th century Henry Darcy developed what is now known as Darcy's Law describing the flow of fluids in porous media. Joseph Boussinesq (a mathematician and physicist) developed theories of stress distribution in elastic solids that proved useful for estimating stresses at depth in the ground; William Rankine, an engineer and physicist, developed an alternative to Coulomb's earth pressure theory. Albert Atterberg developed the clay consistency indices that are still used today for soil classification. Osborne Reynolds recognized in 1885 that shearing causes volumetric dilation of dense and contraction of loose granular materials.

Modern geotechnical engineering is said to have begun in 1925 with the publication of *Erdbaumechanik* by Karl Terzaghi (a civil engineer and geologist). Considered by many to be the father of modern soil mechanics and geotechnical engineering, Terzaghi developed the principle of effective stress, and demonstrated that the shear strength of soil is controlled by effective stress. Terzaghi also developed the framework for theories of bearing capacity of foundations, and the theory for prediction of the rate of settlement of clay layers due to consolidation. In his 1948 book, Donald Taylor recognized that interlocking and dilation of densely packed particles contributed to the peak strength of a soil. The interrelationships between volume change behavior (dilation, contraction, and consolidation) and shearing behavior were all connected via the theory of plasticity using critical state soil mechanics by Roscoe, Schofield, and Wroth with the publication of "On the Yielding of Soils" in 1958. Critical state soil mechanics is the basis for many contemporary advanced constitutive models describing the behavior of soil.

Geotechnical centrifuge modeling is a method of testing physical scale models of geotechnical problems. The use of a centrifuge enhances the similarity of the scale model tests involving soil because the strength and stiffness of soil is very sensitive to the confining pressure. The centrifugal acceleration allows a researcher to obtain large (prototype-scale) stresses in small physical models.

Practicing Engineers

Geotechnical engineers are typically graduates of a four-year civil engineering program and some hold a masters degree. In the USA, geotechnical engineers are typically licensed and regulated as Professional Engineers (PEs) in most states; currently only California and Oregon have licensed geotechnical engineering specialties. The Academy of Geo-Professionals (AGP) began issuing Diplomate, Geotechnical Engineering (D.GE) certification in 2008. State governments will typically license engineers who have graduated from an ABET accredited school, passed the Fundamentals of Engineering examination, completed several years of work experience under the supervision of a licensed Professional Engineer, and passed the Professional Engineering examination.

Soil Mechanics

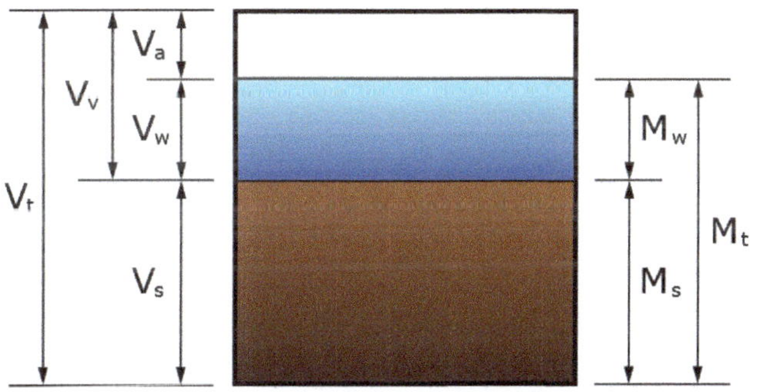

A phase diagram of soil indicating the weights and volumes of air, soil, water, and voids.

In geotechnical engineering, soils are considered a three-phase material composed of: rock or mineral particles, water and air. The voids of a soil, the spaces in between mineral particles, contain the water and air.

The engineering properties of soils are affected by four main factors: the predominant size of the mineral particles, the type of mineral particles, the grain size distribution, and the relative quantities of mineral, water and air present in the soil matrix. Fine particles (fines) are defined as particles less than 0.075 mm in diameter.

Soil Properties

Some of the important properties of soils that are used by geotechnical engineers to analyze site conditions and design earthworks, retaining structures, and foundations are:

Specific weight or Unit Weight

> Cumulative weight of the solid particles, water and air of the unit volume of soil. Note that the air phase is often assumed to be weightless.

Porosity

Ratio of the volume of voids (containing air, water, or other fluids) in a soil to the total volume of the soil. Porosity is mathematically related to void ratio the by

$$n = \frac{e}{1+e}$$

here e is void ratio and n is porosity

Void ratio

The ratio of the volume of voids to the volume of solid particles in a soil mass. Void ratio is mathematically related to the porosity by

$$e = \frac{n}{1-n}$$

Permeability

A measure of the ability of water to flow through the soil. It is expressed in units of velocity.

Compressibility

The rate of change of volume with effective stress. If the pores are filled with water, then the water must be squeezed out of the pores to allow volumetric compression of the soil; this process is called consolidation.

Shear strength

The maximum shear stress that can be applied in a soil mass without causing shear failure.

Atterberg Limits

Liquid limit, Plastic limit, and Shrinkage limit. These indices are used for estimation of other engineering properties and for soil classification.

Geotechnical Investigation

Geotechnical engineers and engineering geologists perform geotechnical investigations to obtain information on the physical properties of soil and rock underlying (and sometimes adjacent to) a site to design earthworks and foundations for proposed structures, and for repair of distress to earthworks and structures caused by subsurface conditions. A geotechnical investigation will include surface exploration and subsurface exploration of a site. Sometimes, geophysical methods are used to obtain data about sites. Subsurface exploration usually involves in-situ testing (two common examples of in-situ tests are the standard penetration test and cone penetration test). In addition site investigation will often include subsurface sampling and laboratory testing of the soil samples retrieved. The digging of test pits and trenching (particularly for locating faults and slide planes) may also be used to learn about soil conditions at depth. Large diameter borings are rarely used due to safety concerns and expense, but are sometimes used to allow a geologist or engineer to be lowered into the borehole for direct visual and manual examination of the soil and rock stratigraphy.

A variety of soil samplers exist to meet the needs of different engineering projects. The standard penetration test (SPT), which uses a thick-walled split spoon sampler, is the most common way to collect disturbed samples. Piston samplers, employing a thin-walled tube, are most commonly used for the collection of less disturbed samples. More advanced methods, such as ground freezing and the Sherbrooke block sampler, are superior, but even more expensive.

Atterberg limits tests, water content measurements, and grain size analysis, for example, may be performed on disturbed samples obtained from thick walled soil samplers. Properties such as shear strength, stiffness hydraulic conductivity, and coefficient of consolidation may be significantly altered by sample disturbance. To measure these properties in the laboratory, high quality sampling is required. Common tests to measure the strength and stiffness include the triaxial shear and unconfined compression test.

Surface exploration can include geologic mapping, geophysical methods, and photogrammetry; or it can be as simple as an engineer walking around to observe the physical conditions at the site. Geologic mapping and interpretation of geomorphology is typically completed in consultation with a geologist or engineering geologist.

Geophysical exploration is also sometimes used. Geophysical techniques used for subsurface exploration include measurement of seismic waves (pressure, shear, and Rayleigh waves), surface-wave methods and/or downhole methods, and electromagnetic surveys (magnetometer, resistivity, and ground-penetrating radar).

Foundations

A building's foundation transmits loads from buildings and other structures to the earth. Geotechnical engineers design foundations based on the load characteristics of the structure and the properties of the soils and/or bedrock at the site. In general, geotechnical engineers:

1. Estimate the magnitude and location of the loads to be supported.
2. Develop an investigation plan to explore the subsurface.
3. Determine necessary soil parameters through field and lab testing (e.g., consolidation test, triaxial shear test, vane shear test, standard penetration test).

4. Design the foundation in the safest and most economical manner.

The primary considerations for foundation support are bearing capacity, settlement, and ground movement beneath the foundations. Bearing capacity is the ability of the site soils to support the loads imposed by buildings or structures. Settlement occurs under all foundations in all soil conditions, though lightly loaded structures or rock sites may experience negligible settlements. For heavier structures or softer sites, both overall settlement relative to unbuilt areas or neighboring buildings, and differential settlement under a single structure, can be concerns. Of particular concern is settlement which occurs over time, as immediate settlement can usually be compensated for during construction. Ground movement beneath a structure's foundations can occur due to shrinkage or swell of expansive soils due to climatic changes, frost expansion of soil, melting of permafrost, slope instability, or other causes. All these factors must be considered during design of foundations.

Many building codes specify basic foundation design parameters for simple conditions, frequently varying by jurisdiction, but such design techniques are normally limited to certain types of construction and certain types of sites, and are frequently very conservative.

In areas of shallow bedrock, most foundations may bear directly on bedrock; in other areas, the soil may provide sufficient strength for the support of structures. In areas of deeper bedrock with soft overlying soils, deep foundations are used to support structures directly on the bedrock; in areas where bedrock is not economically available, stiff "bearing layers" are used to support deep foundations instead.

Shallow Foundations

Example of a slab-on-grade foundation.

Shallow foundations are a type of foundation that transfers building load to the very near the surface, rather than to a subsurface layer. Shallow foundations typically have a depth to width ratio of less than 1.

Footings

Footings (often called "spread footings" because they spread the load) are structural elements which transfer structure loads to the ground by direct areal contact. Footings can be isolated footings for point or column loads, or strip footings for wall or other long (line) loads. Footings are normally constructed from reinforced concrete cast directly onto the soil, and are typically embedded into the ground to penetrate through the zone of frost movement and/or to obtain additional bearing capacity.

Slab Foundations

A variant on spread footings is to have the entire structure bear on a single slab of concrete underlying the entire area of the structure. Slabs must be thick enough to provide sufficient rigidity to spread the bearing loads somewhat uniformly, and to minimize differential settlement across the foundation. In some cases, flexure is allowed and the building is constructed to tolerate small movements of the foundation instead. For small structures, like single-family houses, the slab may be less than 300 mm thick; for larger structures, the foundation slab may be several meters thick.

Slab foundations can be either slab-on-grade foundations or embedded foundations, typically in buildings with basements. Slab-on-grade foundations must be designed to allow for potential ground movement due to changing soil conditions.

Deep Foundations

Pile-driving for a bridge in Napa, California.

Deep foundations are used for structures or heavy loads when shallow foundations cannot provide adequate capacity, due to size and structural limitations. They may also be used to transfer building loads past weak or compressible soil layers. While shallow foundations rely solely on the bearing capacity of the soil beneath them, deep foundations can rely on end bearing resistance, frictional resistance along their length, or both in developing the required capacity. Geotechnical engineers use specialized tools, such as the cone penetration test, to estimate the amount of skin and end bearing resistance available in the subsurface.

There are many types of deep foundations including piles, drilled shafts, caissons, piers, and earth stabilized columns. Large buildings such as skyscrapers typically require deep foundations. For example, the Jin Mao Tower in China uses tubular steel piles about 1m (3.3 feet) driven to a depth of 83.5m (274 feet) to support its weight.

In buildings that are constructed and found to undergo settlement, underpinning piles can be used to stabilise the existing building.

There are three ways to place piles for a deep foundation. They can be driven, drilled, or installed by use of an auger. Driven piles are extended to their necessary depths with the application of external energy in the same way a nail is hammered. There are four typical hammers used to drive such piles: drop hammers, diesel hammers, hydraulic hammers, and air hammers. Drop hammers simply drop a heavy weight onto the pile to drive it, while diesel hammers use a single cylinder diesel engine to force piles through the Earth. Similarly, hydraulic and air hammers supply energy to piles through hydraulic and air forces. Energy imparted from a hammer head varies with type of hammer chosen, and can be as high as a million foot pounds for large scale diesel hammers, a very common hammer head used in practice. Piles are made of a variety of material including steel, timber, and concrete. Drilled piles are created by first drilling a hole to the appropriate depth, and filling it with concrete. Drilled piles can typically carry more load than driven piles, simply due to a larger diameter pile. The auger method of pile installation is similar to drilled pile installation, but concrete is pumped into the hole as the auger is being removed.

Lateral Earth Support Structures

A retaining wall is a structure that holds back earth. Retaining walls stabilize soil and rock from downslope movement or erosion and provide support for vertical or near-vertical grade changes. Cofferdams and bulkheads, structures to hold back water, are sometimes also considered retaining walls.

The primary geotechnical concern in design and installation of retaining walls is that the weight of the retained material is creates lateral earth pressure behind the wall, which can cause the wall to deform or fail. The lateral earth pressure depends on the height of the wall, the density of the soil, the strength of the soil, and the amount of allowable movement of the wall. This pressure is smallest at the top and increases toward the bottom in a manner similar to hydraulic pressure, and tends to push the wall away from the backfill. Groundwater behind the wall that is not dissipated by a drainage system causes an additional horizontal hydraulic pressure on the wall.

Gravity Walls

Gravity walls depend on the size and weight of the wall mass to resist pressures from behind. Gravity walls will often have a slight setback, or batter, to improve wall stability. For short, landscaping walls, gravity walls made from dry-stacked (mortarless) stone or segmental concrete units (masonry units) are commonly used.

Earlier in the 20th century, taller retaining walls were often gravity walls made from large masses of concrete or stone. Today, taller retaining walls are increasingly built as composite gravity walls such as: geosynthetic or steel-reinforced backfill soil with precast facing; gabions (stacked steel wire baskets filled with rocks), crib walls (cells built up log cabin style from precast concrete or timber and filled with soil or free draining gravel) or soil-nailed walls (soil reinforced in place with steel and concrete rods).

For reinforced-soil gravity walls, the soil reinforcement is placed in horizontal layers throughout the height of the wall. Commonly, the soil reinforcement is geogrid, a high-strength polymer mesh, that provide tensile strength to hold soil together. The wall face is often of precast, segmental concrete units that can tolerate some differential movement. The reinforced soil's mass, along with the facing, becomes the gravity wall. The reinforced mass must be built large enough to retain the pressures from the soil behind it. Gravity walls usually must be a minimum of 30 to 40 percent as deep (thick) as the height of the wall, and may have to be larger if there is a slope or surcharge on the wall.

Cantilever Walls

Prior to the introduction of modern reinforced-soil gravity walls, cantilevered walls were the most common type of taller retaining wall. Cantilevered walls are made from a relatively thin stem of steel-reinforced, cast-in-place concrete or mortared masonry (often in the shape of an inverted T). These walls cantilever loads (like a beam) to a large, structural footing; converting horizontal pressures from behind the wall to vertical pressures on the ground below. Sometimes cantilevered walls are buttressed on the front, or include a counterfort on the back, to improve their stability against high loads. Buttresses are short wing walls at right angles to the main trend of the wall. These walls require rigid concrete footings below seasonal frost depth. This type of wall uses much less material than a traditional gravity wall.

Cantilever walls resist lateral pressures by friction at the base of the wall and/or passive earth pressure, the tendency of the soil to resist lateral movement.

Basements are a form of cantilever walls, but the forces on the basement walls are greater than on conventional walls because the basement wall is not free to move.

Excavation Shoring

Shoring of temporary excavations frequently requires a wall design which does not extend laterally beyond the wall, so shoring extends below the planned base of the excavation. Common methods of shoring are the use of sheet piles or soldier beams and lagging. Sheet piles are a form of driven piling using thin interlocking sheets of steel to obtain a continuous barrier in the ground, and are driven prior to excavation. Soldier beams are constructed of wide flange steel H sections spaced about 2–3 m apart, driven prior to excavation. As the excavation proceeds, horizontal timber or steel sheeting (lagging) is inserted behind the H pile flanges.

In some cases, the lateral support which can be provided by the shoring wall alone is insufficient to resist the planned lateral loads; in this case additional support is provided by walers or tie-backs. Walers are structural elements which connect across the excavation so that the loads from the soil on either side of the excavation are used to resist each other, or which transfer horizontal loads from the shoring wall to the base of the excavation. Tie-backs are steel tendons drilled into the face of the wall which extend beyond the soil which is applying pressure to the wall, to provide additional lateral resistance to the wall.

Earthworks

Excavation

Excavation is the process of training earth according to requirement by removing the soil from the site.

Filling

Filling is the process of training earth according to requirement by placing the soil on the site.

Compaction

Compaction is the process by which the density of soil is increased and permeability of soil is decreased. Fill placement work often has specifications requiring a specific degree of compaction, or alternatively, specific properties of the compacted soil. In-situ soils can be compacted by rolling, deep dynamic compaction, vibration, blasting, gyrating, kneading, compaction grouting etc.

A compactor/roller operated by U.S. Navy Seabees

Ground Improvement

Ground Improvement is a technique that improves the engineering properties of the treated soil mass. Usually, the properties modified are shear strength, stiffness and permeability. Ground improvement has developed into a sophisticated tool to support foundations for a wide variety of structures. Properly applied, i.e. after giving due consideration to the nature of the ground being improved and the type and sensitivity of the structures being built, ground improvement often reduces direct costs and saves time.

Slope Stabilization

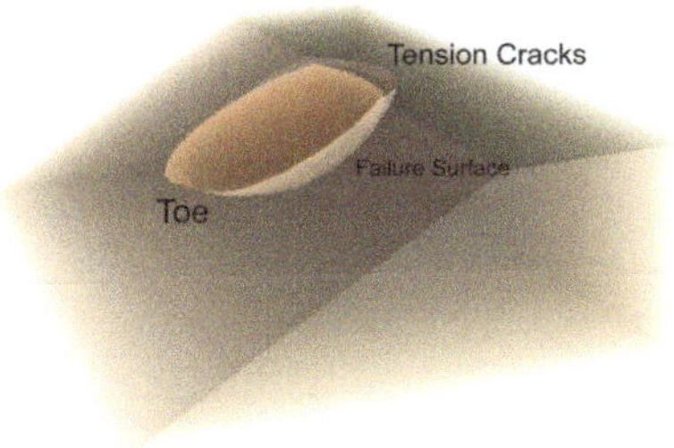

Simple slope slip section.

Slope stability is the potential of soil covered slopes to withstand and undergo movement. Stability is determined by the balance of shear stress and shear strength. A previously stable slope may be initially affected by preparatory factors, making the slope conditionally unstable. Triggering factors of a slope failure can be climatic events can then make a slope actively unstable, leading to mass movements. Mass movements can be caused by increases in shear stress, such as loading, lateral pressure, and transient forces. Alternatively, shear strength may be decreased by weathering, changes in pore water pressure, and organic material.

Several modes of failure for earth slopes include falls, topples, slides, and flows. In slopes with coarse grained soil or rocks, falls typically occur as the rapid descent of rocks and other loose slope material. A slope topples when a large column of soil tilts over its vertical axis at failure. Typical slope stability analysis considers sliding failures, categorized mainly as rotational slides or translational slides. As implied by the name, rotational slides fail along a generally curved surface, while translational slides fail along a more planar surface. A slope failing as a flow would resemble a fluid flowing downhill.

Slope Stability Analysis

Stability analysis is needed for the design of engineered slopes and for estimating the risk of slope failure in natural or designed slopes. A common assumption is that a slope consists of a layer of soil sitting on top of a rigid base. The mass and the base are assumed to interact via friction. The interface between the mass and the base can be planar, curved, or have some other complex geometry. The goal of a slope stability analysis is to determine the conditions under which the mass will slip relative to the base and lead to slope failure.

If the interface between the mass and the base of a slope has a complex geometry, slope stability analysis is difficult and numerical solution methods are required. Typically, the exact geometry of the interface is not known and a simplified interface geometry is assumed. Finite slopes require three-dimensional models to be analyzed. To keep the problem simple, most slopes are analyzed assuming that the slopes are infinitely wide and can therefore be represented by two-dimensional models. A slope can be drained or undrained. The undrained condition is used in the calculations to produce conservative estimates of risk.

A popular stability analysis approach is based on principles pertaining to the limit equilibrium concept. This method analyzes a finite or infinite slope as if it were about to fail along its sliding failure surface. Equilibrium stresses are calculated along the failure plane, and compared to the soils shear strength as determined by Terzaghi's shear strength equation. Stability is ultimately decided by a factor of safety equal to the ratio of shear strength to the equilibrium stresses along the failure surface. A factor of safety greater than one generally implies a stable slope, failure of which should not occur assuming the slope is undisturbed. A factor of safety of 1.5 for static conditions is commonly used in practice.

Offshore Geotechnical Engineering

Platforms offshore Mexico.

Offshore (or *marine*) *geotechnical engineering* is concerned with foundation design for human-made structures in the sea, away from the coastline (in opposition to *onshore* or *nearshore*). Oil platforms, artificial islands and submarine pipelines are examples of such structures. There are number of significant differences between onshore and offshore geotechnical engineering. Notably, ground improvement (on the seabed) and site investigation are more expensive, the offshore structures are exposed to a wider range of geohazards, and the environmental and financial consequences are higher in case of failure. Offshore structures are exposed to various environmental loads, notably wind, waves and currents. These phenomena may affect the integrity or the serviceability of the structure and its foundation during its operational lifespan – they need to be taken into account in offshore design.

In subsea geotechnical engineering, seabed materials are considered a two-phase material composed of 1) rock or mineral particles and 2) water. Structures may be fixed in place in the seabed—as is the case for piers, jettys and fixed-bottom wind turbines—or may be a floating structure that remain roughly fixed relative to its geotechnical anchor point. Undersea mooring of human-engineered floating structures include a large number of offshore oil and gas platforms and, since 2008, a few floating wind turbines. Two common types of engineered design for anchoring floating structures include tension-leg and catenary loose mooring systems. "Tension leg mooring systems have vertical tethers under tension providing large restoring moments in pitch and roll. Catenary mooring systems provide station keeping for an offshore structure yet provide little stiffness at low tensions."

Geosynthetics

Geosynthetics are a type of plastic polymer products used in geotechnical engineering that improve engineering performance while reducing costs. This includes geotextiles, geogrids, geomembranes, geocells, and geocomposites.

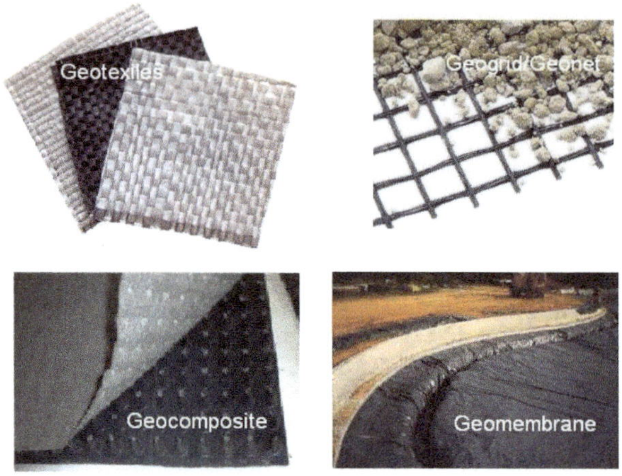

A collage of geosynthetic products.

The synthetic nature of the products make them suitable for use in the ground where high levels of durability are required; their main functions include: drainage, filtration, reinforcement, separation and containment. Geosynthetics are available in a wide range of forms and materials, each to suit a slightly different end use, although they are frequently used together. These products have a wide range of applications and are currently used in many civil and geotechnical engineering applications including: roads, airfields, railroads, embankments, piled embankments, retaining structures, reservoirs, canals, dams, landfills, bank protection and coastal engineering.

Coastal Engineering

Coastal engineering is the branch of civil engineering concerning the specific demands posed by constructing at or near the coast, as well as the development of the coast itself.

Wave attack on Ilfracombe's sea walls during a storm.

The hydrodynamic impact of especially waves, tides, storm surges and tsunamis and (often) the harsh environment of salt seawater are typical challenges for the coastal engineer – as are the morphodynamic changes of the coastal topography, caused both by the autonomous development of the system and man-made changes. The areas of interest in coastal engineering include the coasts of the oceans, seas, marginal seas, estuaries and big lakes.

Besides the design, building and maintenance of coastal structures, coastal engineers are often interdisciplinary involved in integrated coastal zone management, also because of their specific knowledge of the hydro- and morphodynamics of the coastal system. This may include providing input and technology for e.g. environmental impact assessment, port development, strategies for coastal defense, land reclamation, offshore wind farms and other energy-production facilities, etc.

Specific Elements of Coastal Engineering

Beach nourishment at the Dutch coast.

The coastal environment produces challenges specific for this branch of engineering: waves, storm surges, tides, tsunamis, sea level changes, sea water and the marine ecosystem.

Most often, in coastal engineering projects there is a need for metocean conditions: local wind and wave climate, as well as statistics for and information on other hydrodynamic quantities of interest. Also, bathymetry and morphological changes are of direct interest. In case of studies of sediment transport and morphological changes, relevant properties of the sea bed sediments, water and ecosystem properties are needed.

Long and Short Waves

The occurrence of wave phenomena – like sea waves, swell, tides and tsunamis – require engineering knowledge of their physics, as well as models: both numerical models and physical models. The practices in present-day coastal engineering are more-and-more based on models verified and validated by experimental data.

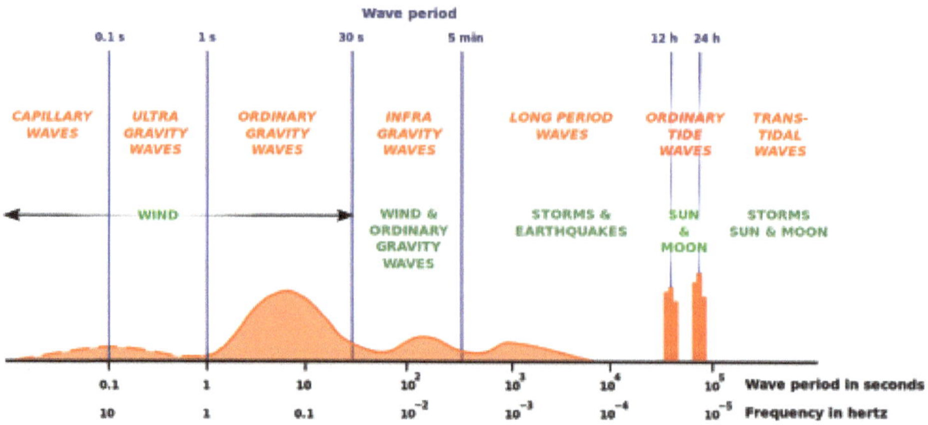

Classification of the spectrum of ocean waves according to wave period, by Walter Munk (1950).

Apart from the wave transformations themselves, for the waves coming from deep water into the shallow coastal waters and surf zone, the effects of the waves are important. These effects include:

- the wave loading on coastal structures like breakwaters, groynes, jetties, sea walls and dikes
- wave-induced currents, like the longshore current in the surf zone, rip currents and Stokes drift, affecting sediment transport and morphodynamics
- wave agitation in harbors, which may result in harbor downtime
- wave overtopping over seawalls and dikes, which may e.g. threaten the stability of a dike

Facade Engineering

Building facades are one of the largest, most important elements in the overall aesthetic and technical performance of a building. Façade Engineering is the art and science of resolving aesthetic, environmental and structural issues to achieve the effective enclosure of buildings.

Specialist companies are dedicated to this niche sector of the building industry and engineers operate within technical divisions of façade manufacturing companies. Generally, façade engineers are specifically qualified in the discipline of façade engineering and consultants work with the design team on construction projects for architects, building owners, construction managers and product manufacturers.

Façade engineers must consider aspects such as the design, certification, fabrication and installation of the building facades with regards to the performance of materials, aesthetic appearance, structural behaviour, weathertightness, safety and serviceability, security, maintenance and build ability. The skill set will include matters such as computational fluid dynamics, heat transfer through two- and three-dimensional constructions, the behaviour of materials, manufacturing methodologies, structural engineering and logistics.

Over time, the specialist skills necessary in this niche sector have surpassed the capabilities of architects, structural and mechanical engineers as buildings are designed with more complexity and with the introduction of Building Information Modelling (BIM).

Building facades are considered to be one of the most expensive and potentially the highest risk element of any major project. Historically building facades have the greatest level of failure of any part of a building fabric and the pressure for change and adaptation due to environmental and energy performance needs is greater than any other element of a building. As a consequence façade engineering has become a science in its own right.

In the United Kingdom, a professional body associated with the industry is the Society of Façade Engineering. Qualifications in Façade Engineering recognised by the Society of Façade Engineering and international professional qualifications include the MSc in Façade Engineering. This may be from the University of Bath; Technical University Delft or Detmolder Schule fur Architektur und Innenarchitekter Hochschule or other qualifications subject to review by the Membership panel.

References

- *Bozorgnia, Yousef; Bertero, Vitelmo V. (2004). Earthquake Engineering: From Engineering Seismology to Performance-Based Engineering. CRC Press. ISBN 978-0-8493-1439-1.*
- *Newmark, N.M.; Hall, W.J. (1982). Earthquake Spectra and Design. EERI. ISBN 0-943198-22-4.*
- *Clough, Ray W.; Penzien, Joseph (1993). Dynamics of Structures. McGraw-Hill. ISBN 0-07-011394-7*
- *Lindeburg, Michael R.; Baradar, Majid (2001). Seismic Design of Building Structures. Professional Publications. ISBN 1-888577-52-5.*
- *Chu, S.Y.; Soong, T.T.; Reinhorn, A.M. (2005). Active, Hybrid and Semi-Active Structural Control. John Wiley & Sons. ISBN 0-470-01352-4.*
- *Housner, George W.; Jennings, Paul C. (1982). Earthquake Design Criteria. EERI. ISBN 1-888577-52-5.*
- *Arnold, Christopher; Reitherman, Robert (1982). Building Configuration & Seismic Design. A Wiley-Interscience Publication. ISBN 0-471-86138-3.*
- *Robert W. Day (2007). Geotechnical Earthquake Engineering Handbook. McGraw Hill. ISBN 0-07-137782-4.*
- *Edited by Robert Lark (2007). Bridge Design, Construction and Maintenance. Thomas Telford. ISBN 0-7277-3593-4.*

- *Timber Design & Construction Sourcebook=Gotz, Karl-Heinz et al. McGraw-Hall. 1989. ISBN 0-07-023851-0.*
- *Nilson, Arthur H. (1987). Design of Prestressed Concrete. John Wiley & Sons. ISBN 0-471-83072-0.*
- *Reitherman, Robert (2012). Earthquakes and Engineers: An International History. Reston, VA: ASCE Press. pp. 394–395. ISBN 9780784410714.*
- *Seismic Design Manual. Chicago: American Institute of Steel Construction. 2006. pp. 6.1–30. ISBN 1-56424-056-8.*
- *EERI Endowment Subcommittee (May 2000). Financial Management of Earthquake Risk. EERI Publication. ISBN 0-943198-21-6.*
- *Craig Taylor; Erik VanMarcke, eds. (2002).* Acceptable Risk Processes: Lifeline and Natural Hazards. *Reston, VA: ASCE, TCLEE.* ISBN 9780784406236.

3

Important Aspects and Fields of Civil Engineering

Civil engineering is not only applied for environmental protection or social welfare. It has a plethora of commercial applications as well. This chapter will focus on such applications of civil engineering. The topics included in this chapter are transportation engineering, coastal management, construction surveying, dry dock, etc.

Construction

In large construction projects, such as this skyscraper in Melbourne, Australia, cranes are essential.

Construction is the process of constructing a building or infrastructure. Construction differs from manufacturing in that manufacturing typically involves mass production of similar items without a designated purchaser, while construction typically takes place on location for a known client. Construction as an industry comprises six to nine percent of the gross domestic product of developed countries. Construction starts with planning, design, and financing; and continues until the project is built and ready for use.

Large-scale construction requires collaboration across multiple disciplines. An architect normally manages the job, and a construction manager, design engineer, construction engineer or project manager supervises it. For the successful execution of a project, effective planning is essential. Those involved with the design and execution of the infrastructure in question must consider zoning requirements, the environmental impact of the job, the successful scheduling, budgeting, construction-site safety, availability and transportation of building materials, logistics, inconvenience to the public caused by construction delays and bidding, etc. The largest construction projects are referred to as megaprojects.

Etymology

Construction is a general term meaning the art and science to form objects, systems, or organizations, and comes from Latin *constructionem* (from *com-* "together" and *struere* "to pile up") and Old French *construction*. Construction is used as a verb: the act of building, and a noun: how a building was built, the nature of its structure.

Types of Construction

Military residential unit construction by U.S. Navy personnel in Afghanistan

In general, there are three sectors of construction: buildings, infrastructure and industrial. Building construction is usually further divided into residential and non-residential (commercial/institutional). Infrastructure is often called heavy/highway, heavy civil or heavy engineering. It includes large public works, dams, bridges, highways, water/wastewater and utility distribution. Industrial includes refineries, process chemical, power generation, mills and manufacturing plants. There are other ways to break the industry into sectors or markets.

Engineering News-Record (ENR) is a trade magazine for the construction industry. Each year, ENR compiles and reports on data about the size of design and construction companies. They publish a list of the largest companies in the United States (Top-40) and also a list the largest global firms (Top-250, by amount of work they are doing outside their home country). In 2014, ENR compiled the data in nine market segments. It was divided as transportation, petroleum, buildings, power, industrial, water, manufacturing, sewer/waste, telecom, hazardous waste plus a tenth category for other projects. In their reporting on the Top 400, they used data on transportation, sewer, hazardous waste and water to rank firms as heavy contractors.

The Standard Industrial Classification and the newer North American Industry Classification System have a classification system for companies that perform or otherwise engage in construction. To recognize the differences of companies in this sector, it is divided into three subsectors: building construction, heavy and civil engineering construction, and specialty trade contractors. There are also categories for construction service firms (e.g., engineering, architecture) and construction managers (firms engaged in managing construction projects without assuming direct financial responsibility for completion of the construction project).

Building Construction

Building construction is the process of adding structure to real property or construction of buildings. The majority of building construction jobs are small renovations, such as addition of a room, or renovation of a bathroom. Often, the owner of the property acts as laborer, paymaster, and design team for the entire project. Although building construction projects typically include various common elements, such as design, financial, estimating and legal considerations, many projects of varying sizes reach undesirable end results, such as structural collapse, cost overruns, and/or litigation. For this reason, those with experience in the field make detailed plans and maintain careful oversight during the project to ensure a positive outcome.

The National Cement Share Company of Ethiopia's new plant in Dire Dawa.

Commercial building construction is procured privately or publicly utilizing various delivery methodologies, including cost estimating, hard bid, negotiated price, traditional, management contracting, construction management-at-risk, design & build and design-build bridging.

Residential construction practices, technologies, and resources must conform to local building authority regulations and codes of practice. Materials readily available in the area generally dictate the construction materials used (e.g. brick versus stone, versus timber). Cost of construction on a per square meter (or per square foot) basis for houses can vary dramatically based on site conditions, local regulations, economies of scale (custom designed homes are often more expensive to build) and the availability of skilled tradespeople. As residential construction (as well as all other types of construction) can generate a lot of waste, careful planning again is needed here.

Residential Construction

The most popular method of residential construction in North America is wood-framed construction. Typical construction steps for a single-family or small multi-family house are:

- Develop floor plans and obtain a materials list for estimations (more recently performed with estimating software)
- Obtain government building approval if necessary
- Clear the building site
- Survey to stake out for the foundation
- Excavate the foundation and dig footers.
- Pour a foundation and footers with concrete
- Build the main load-bearing structure out of thick pieces of wood and possibly metal I-beams for large spans with few supports.
- Add floor and ceiling joists and install subfloor panels
- Cover outer walls and roof in OSB or plywood and a water-resistive barrier.
- Install roof shingles or other covering for flat roof
- Cover the walls with siding, typically vinyl, wood, or brick veneer but possibly stone or other materials
- Install windows
- Frame interior walls with wooden 2x4s
- Add internal plumbing, HVAC, electrical, and natural gas utilities

Framing

- Building inspector visits if necessary to approve utilities and framing
- Install insulation and interior drywall panels (cementboard for wet areas) and to complete walls and ceilings
- Install bathroom fixtures
- Spackle, prime, and paint interior walls and ceilings
- Additional tiling on top of cementboard for wet areas, such as the bathroom and kitchen backsplash
- Install final floor covering, such as floor tile, carpet, or wood flooring
- Install major appliances
- Unless the original owners are building the house, at this point it is typically sold or rented.

New Construction Techniques and Sustainability

As efficiency codes have come into effect in recent years, new construction technologies and methods have emerged. University Construction Management departments are on the cutting edge of the newest methods of construction intended to improve efficiency, performance and reduce construction waste.

New techniques of building construction are being researched, made possible by advances in 3D printing technology. In a form of *additive building construction*, similar to the additive manufacturing techniques for manufactured parts, building printing is making it possible to flexibly construct small commercial buildings and private habitations in around 20 hours, with built-in plumbing and electrical facilities, in one continuous build, using large 3D printers. Working versions of 3D-printing building technology are already printing 2 metres (6 ft 7 in) of building material per hour as of January 2013, with the next-generation printers capable of 3.5 metres (11 ft) per hour, sufficient to complete a building in a week. Dutch architect Janjaap Ruijssenaars's performative architecture 3D-printed building is scheduled to be built in 2014.

In the current trend of sustainable construction, the recent movements of New Urbanism and New Classical Architecture promote a sustainable approach towards construction, that appreciates and develops smart growth, architectural tradition and classical design. This is in contrast to modernist and short-lived globally uniform architecture, as well as opposing solitary housing estates and suburban sprawl. Both trends started in the 1980s.

The construction site may be shut down due to bad weather. Erecting scaffolded tents over the site may reduce the number of lost work days, increasing productivity.

Construction Processes

Design Team

Shasta Dam under construction in June 1942

In the modern industrialized world, construction usually involves the translation of designs into reality. A formal design team may be assembled to plan the physical proceedings, and to integrate those proceedings with the other parts. The design usually consists of drawings and specifications, usually prepared by a design team including Architect, civil engineers, mechanical engineers, electrical engineers, structural engineers, fire protection engineers, planning consultants, architectural consultants, and archaeological consultants. The design team is most commonly employed by (i.e. in contract with) the property owner. Under this system, once the design is completed by the design team, a number of construction companies or construction management companies may then be asked to make a bid for the work, either based directly on the design, or on the basis of drawings and a bill of quantities provided by a quantity surveyor. Following evaluation of bids, the owner typically awards a contract to the most cost efficient bidder.

The modern trend in design is toward integration of previously separated specialties, especially among large firms. In the past, architects, interior designers, engineers, developers, construction managers, and general contractors were more likely to be entirely separate companies, even in the larger firms. Presently, a firm that is nominally an "architecture" or "construction management" firm may have experts from all related fields as employees, or to have an associated company that provides each necessary skill. Thus, each such firm may offer itself as "one-stop shopping" for a construction project, from beginning to end. This is designated as a "design build" contract where the contractor is given a performance specification and must undertake the project from design to construction, while adhering to the performance specifications.

Several project structures can assist the owner in this integration, including design-build, partnering and construction management. In general, each of these project structures allows the owner to integrate the services of architects, interior designers, engineers and constructors throughout design and construction. In response, many companies are growing beyond traditional offerings of design or construction services alone and are placing more emphasis on establishing relationships with other necessary participants through the design-build process.

The increasing complexity of construction projects creates the need for design professionals trained in all phases of the project's life-cycle and develop an appreciation of the building as an advanced technological system requiring close integration of many sub-systems and their individual components, including sustainability. Building engineering is an emerging discipline that attempts to meet this new challenge.

Financial Advisors

Trump International Hotel and Tower (Chicago)

Construction projects can suffer from preventable financial problems. Underbids happen when builders ask for too little money to complete the project. Cash flow problems exist when the present amount of funding cannot cover the current costs for labour and materials, and because they are a matter of having sufficient funds at a specific time, can arise even when the overall total is enough. Fraud is a problem in many fields, but is notoriously prevalent in the construction field. Financial planning for the project is intended to ensure that a solid plan with adequate safeguards and contingency plans are in place before the project is started and is required to ensure that the plan is properly executed over the life of the project.

May 23, 2006

September 14, 2007 (3 months before completion)

Mortgage bankers, accountants, and cost engineers are likely participants in creating an overall plan for the financial management of the building construction project. The presence of the mortgage banker is highly likely, even in relatively small projects since the owner's equity in the property is the most obvious source of funding for a building project. Accountants act to study the expected monetary flow over the life of the project and to monitor the payouts throughout the process. Cost engineers and estimators apply expertise to relate the work and materials involved to a proper valuation. Cost overruns with government projects have occurred when the contractor identified change orders or project changes that increased costs, which are not subject to competition from other firms as they have already been eliminated from consideration after the initial bid.

Large projects can involve highly complex financial plans and often start with a conceptual estimate performed by a building estimator. As portions of a project are completed, they may be sold, supplanting one lender or owner for another, while the logistical requirements of having the right trades and materials available for each stage of the building construction project carries forward. In many English-speaking countries, but not the United States, projects typically use quantity surveyors.

Legal Aspects

Construction along Ontario Highway 401, widening the road from six to twelve travel lanes

Construction site with tubes

A construction project must fit into the legal framework governing the property. These include governmental regulations on the use of property,and obligations that are created in the process of construction.

The project must adhere to zoning and building code requirements. Constructing a project that fails to adhere to codes does not benefit the owner. Some legal requirements come from malum in se considerations, or the desire to prevent things that are indisputably bad – bridge collapses or explosions. Other legal requirements come from malum prohibitum considerations, or things that are a matter of custom or expectation, such as isolating businesses to a business district and residences to a residential district. An attorney may seek changes or exemptions in the law that governs the land where the building will be built, either by arguing that a rule is inapplicable (the bridge design will not cause a collapse), or that the custom is no longer needed (acceptance of live-work spaces has grown in the community).

A construction project is a complex net of contracts and other legal obligations, each of which all parties must carefully consider. A contract is the exchange of a set of obligations between two or more parties, but it is not so simple a matter as trying to get the other side to agree to as much as possible in exchange for as little as possible. The time element in construction means that a delay costs money, and in cases of bottlenecks, the delay can be extremely expensive. Thus, the contracts must be designed to ensure that each side is capable of performing the obligations set out. Contracts that set out clear expectations and clear paths to accomplishing those expectations are far more likely to result in the project flowing smoothly, whereas poorly drafted contracts lead to confusion and collapse.

Legal advisors in the beginning of a construction project seek to identify ambiguities and other potential sources of trouble in the contract structure, and to present options for preventing problems. Throughout the process of the project, they work to avoid and resolve conflicts that arise. In each case, the lawyer facilitates an exchange of obligations that matches the reality of the project.

Interaction of Expertise

Apartment complex under construction in Daegu, South Korea

Design, finance, and legal aspects overlap and interrelate. The design must be not only structurally sound and appropriate for the use and location, but must also be financially possible to build, and legal to use. The financial structure must accommodate the need for building the design provided, and must pay amounts that are legally owed. The legal structure must integrate the design into the surrounding legal framework, and enforce the financial consequences of the construction process.

Procurement

Procurement describes the merging of activities undertaken by the client to obtain a building. There are many different methods of construction procurement; however the three most common types of procurement are traditional (design-bid-build), design-build and management contracting.

There is also a growing number of new forms of procurement that involve relationship contracting where the emphasis is on a co-operative relationship between the principal and contractor and other stakeholders within a construction project. New forms include partnering such as Public-Private Partnering (PPPs) aka private finance initiatives (PFIs) and alliances such as "pure" or "project" alliances and "impure" or "strategic" alliances. The focus on co-operation is to ameliorate the many problems that arise from the often highly competitive and adversarial practices within the construction industry.

Traditional

This is the most common method of construction procurement and is well established and recognized. In this arrangement, the architect or engineer acts as the project coordinator. His or her role is to design the works, prepare the specifications and produce construction drawings, administer the contract, tender the works, and manage the works from inception to completion. There are direct contractual links between the architect's client and the main contractor. Any subcontractor has a direct contractual relationship with the main contractor. The procedure continues until the building is ready to occupy.

Design-Build

This approach has become more common in recent years, and involves the client contracting a single entity to both provide a design and to build that design. In some cases, the design-build package can also include finding the site, arranging funding and applying for all necessary statutory consents.

The owner produces a list of requirements for a project, giving an overall view of the project's goals. Several D&B contractors present different ideas about how to accomplish these goals. The owner selects the ideas he or she likes best and hires the appropriate contractor. Often, it is not just one contractor, but a consortium of several contractors working together. Once these have been hired, they begin building the first phase of the project. As they build phase 1, they design phase 2. This is in contrast to a design-bid-build contract, where the project is completely designed by the owner, then bid on, then completed.

Construction of the *Phase-1* (first two towers) of the Havelock City Project, Sri Lanka

Kent Hansen pointed out that state departments of transportation usually use design build contracts as a way of progressing projects when states lack the skills-resources. In such departments, design build contracts are usually employed for very large projects.

Management Procurement Systems

In this arrangement the client plays an active role in the procurement system by entering into separate contracts with the designer (architect or engineer), the construction manager, and individual trade contractors. The client takes on the contractual role, while the construction or project manager provides the active role of managing the separate trade contracts, and ensuring that they complete all work smoothly and effectively together.

Management procurement systems are often used to speed up the procurement processes, allow the client greater flexibility in design variation throughout the contract, give the ability to appoint individual work contractors, separate contractual responsibility on each individual throughout the contract, and to provide greater client control.

Authority Having Jurisdiction

In construction, the authority having jurisdiction (AHJ) is the governmental agency or sub-agency that regulates the construction process. In most cases, this is the municipality where the building is located. However, construction performed for supra-municipal authorities are usually regulated directly by the owning authority, which becomes the AHJ.

Construction on the Federal Reserve building in Kansas City, Missouri

Before the foundation can be dug, contractors are typically required to verify and have existing utility lines marked, either by the utilities themselves or through a company specializing in such services. This lessens the likelihood of damage to the existing electrical, water, sewage, phone, and cable facilities, which could cause outages and potentially hazardous situations. During the construction of a building, the municipal building inspector inspects the building periodically to ensure that the construction adheres to the approved plans and the local building code. Once construction is complete and a final inspection has been passed, an occupancy permit may be issued.

An operating building must remain in compliance with the fire code. The fire code is enforced by the local fire department or a municipal code enforcement office.

Changes made to a building that affect safety, including its use, expansion, structural integrity, and fire protection items, usually require approval of the AHJ for review concerning the building code.

Industry Characteristics

In the United States, the industry in 2014 has around $960 billion in annual revenue according to statistics tracked by the Census Bureau, of which $680 billion is private (split evenly between residential and nonresidential) and the remainder is government. As of 2005, there were about 667,000 firms employing 1 million contractors (200,000 general contractors, 38,000 heavy, and 432,000 specialty); the average contractor employed fewer than 10 employees. As a whole, the industry employed an estimated 5.8 million as of April 2013, with a 13.2% unemployment rate. In the United States, approximately 828,000 women were employed in the construction industry as of 2011.

Careers

Helicopter view of the Atacama Large Millimeter/submillimeter Array (ALMA) Operations Support Facility (OSF) construction site

Ironworkers erecting the steel frame of a new building at Massachusetts General Hospital in Boston

A truck operator at Al Gamil, the largest construction company in Djibouti.

There are many routes to the different careers within the construction industry. These three main tiers are based on educational background and training, which vary by country:

- Unskilled and semi-skilled – General site labor with little or no construction qualifications.
- Skilled – Tradesmen who've served apprenticeships, typically in labor unions, and on-site managers who possess extensive knowledge and experience in their craft or profession.
- Technical and management – Personnel with the greatest educational qualifications, usually graduate degrees, trained to design, manage and instruct the construction process.

Skilled occupations include carpenters, electricians, plumbers, ironworkers, masons, and many other manual crafts, as well as those involved in project management. In the UK these require further education qualifications, often in vocational subject areas. These qualifications are either obtained directly after the completion of compulsory education or through "on the job" apprenticeship training. In the UK, 8500 construction-related apprenticeships were commenced in 2007.

Technical and specialized occupations require more training as a greater technical knowledge is required. These professions also hold more legal responsibility. A short list of the main careers with an outline of the educational requirements are given below:

- Architect – Typically holds 1, undergraduate 3 year degree in architecture + 1, post-graduate 2 year degree (DipArch or BArch) in architecture plus 24 months experience within the industry. To use the title "architect" the individual must be registered on the Architects Registration Board register of Architects.

- Civil engineer – Typically holds a degree in a related subject. The Chartered Engineer qualification is controlled by the Engineering Council, and is often achieved through membership of the Institution of Civil Engineers. A new university graduate must hold a master's degree to become chartered; persons with bachelor's degrees may become an Incorporated Engineer.
- Building services engineer – Often referred to as an "M&E Engineer" typically holds a degree in mechanical or electrical engineering. Chartered Engineer status is governed by the Engineering Council, mainly through the Chartered Institution of Building Services Engineers.
- Project manager – Typically holds a 4-year or greater higher education qualification, but are often also qualified in another field such as architecture, civil engineering or quantity surveying.
- Structural engineer – Typically holds a bachelor's or master's degree in structural engineering. A P.ENG is required from the Professional Engineers Ontario (Canada). New university graduates must hold a master's degree to gain chartered status from the Engineering Council, mainly through the Institution of Structural Engineers (UK).
- Quantity surveyor – Typically holds a bachelor's degree in quantity surveying. Chartered status is gained from the Royal Institution of Chartered Surveyors.
- Civil estimators are professionals who typically have a background in civil engineering, construction project management, or construction supervision.

In 2010 a salary survey revealed the differences in remuneration between different roles, sectors and locations in the construction and built environment industry. The results showed that areas of particularly strong growth in the construction industry, such as the Middle East, yield higher average salaries than in the UK for example. The average earning for a professional in the construction industry in the Middle East, across all sectors, job types and levels of experience, is £42,090, compared to £26,719 in the UK. This trend is not necessarily due to the fact that more affluent roles are available, however, as architects with 14 or more years experience working in the Middle East earn on average £43,389 per annum, compared to £40,000 in the UK. Some construction workers in the US/Canada have made more than $100,000 annually, depending on their trade.

Safety

Construction is one of the most dangerous occupations in the world, incurring more occupational fatalities than any other sector in both the United States and in the European Union. In 2009, the fatal occupational injury rate among construction workers in the United States was nearly three times that for all workers. Falls are one of the most common causes of fatal and non-fatal injuries among construction workers. Proper safety equipment such as harnesses and guardrails and procedures such as securing ladders and inspecting scaffolding can curtail the risk of occupational injuries in the construction industry. Other major causes of fatalities in the construction industry include electrocution, transportation accidents, and trench cave-ins.

At-risk workers without appropriate safety equipment

Roofing requires a very high level of safety

Other safety risks for workers in construction include hearing loss due to high noise exposure, musculoskeletal injury, chemical exposure, and high levels of stress.

History

The first huts and shelters were constructed by hand or with simple tools. As cities grew during the Bronze Age, a class of professional craftsmen, like bricklayers and carpenters, appeared. Occasionally, slaves were used for construction work. In the Middle Ages, these were organized into guilds. In the 19th century, steam-powered machinery appeared, and later diesel- and electric powered vehicles such as cranes, excavators and bulldozers.

A picture of a building under construction in India.

Fast-track construction has been increasingly popular in the 21st century. Some estimates suggest that 40% of construction projects are now fast-track construction.

Construction Phases

- Vision/fantasy/idea - a concept never intended to be built, may be an aesthetic or structural design exercise
- Proposed - a building concept that is under review by the building owner and by government
- Approved - a building concept that will be constructed in the near future. If the proposed building is not approved then the proposal may be amended and re-submitted, or it may be deferred or cancelled.
- Design - the specification of what is to be built in sufficient detail to be used as the basis as a contract between the owner and a contractor
- Procurement - the selection of the contractor or contractors to carry out the construction. This may be by competitive tendering.
- Diversions - before construction can start any services on the site which must be kept operational to serve other adjacent sites must be diverted so they run

outside the footprint of the new building. This can include drainage, water and gas piped services and power and communication cables.

- Under-construction - a fully designed building currently being built
- Ground works - construction work below ground level including the construction of basements and foundations
- Topped-out - a fully designed building where construction has reached the highest point of the building
- Fitting out - installation of the decorative, non-structural elements once the building main structure is complete. This includes painting, ceilings, light fittings etc.
- Commissioning or setting to work - Once the building Mechanical, electrical, plumbing, communications, and building control systems are installed they then need to be tested and adjusted so they deliver the required performance. In modern buildings this can take some time during which little seems to be going on but if this is not done properly then these systems will not deliver their design performance leading to hot and cold spots, spurious alarms, higher energy bills, and systems failing during emergencies
- Substantial Completion / Beneficial Occupancy - a point when the work is sufficiently complete so that the Owner can occupy (Items noted during inspection 'punch list' or 'snag list' may still be corrected)
- Complete/built - a fully designed building that has been fully built, excluding future expansions (punch list items all completed)
- Building Operation - All those day-to-day activities need to ensure the building can be used. In simple buildings this means little more than cleaning but in more complicated buildings this is a large scale operation employing a large team of staff. If they do their job right then you hardly notice them.
- Maintenance - works to ensure the building continues to operate in accordance with its design, including replacing elements which are approaching the end of their useful life
- Repair - replacing building elements which have been damaged or which have failed to restore the building to its as-built state
- Renovation - modification to the building. This can be minor modifications that are carried out while the building is occupied or major works where only the structural elements are kept and the building is out of use for years
- Demolition - destruction of the building which may include the salvage of some elements for reuse elsewhere.

List of Countries by the Largest Output in Construction

List of countries with the largest construction output in 2015	
Economy	**Construction output in 2015 (billions in USD)**
(01) **India**	849
(02) **United States**	599
(03) **China**	569
(04) **Japan**	333
(05) **France**	147
(06) **Germany**	143
(07) **United Kingdom**	131
(08) **Canada**	131
(09) **Australia**	115
(10) **Russia**	111
(11) **Brazil**	109
(12) **Italy**	107
(13) **Spain**	104
(14) **Indonesia**	93
(15) **Mexico**	92
(16) **South Korea**	58
(17) **Turkey**	35
(18) **United Arab Emirates**	34
(19) **Venezuela**	34
(20) **Netherlands**	34
(21) **Poland**	34
(22) **Switzerland**	33
(23) **Saudi Arabia**	32
(24) **Iran**	29
(25) **Colombia**	29
The twenty-five largest countries in the world by construction output (2012)	

Construction Surveying

Construction surveying or building surveying (otherwise known as "staking", "stake-out", "lay-out" or "setting-out") is to stake out reference points and markers that will guide the construction of new structures such as roads or buildings. These markers are usually staked out according to a suitable coordinate system selected for the project.

History of Construction Surveying

- The nearly perfect squareness and north-south orientation of the Great Pyramid of Giza, built c. 2700 BC, affirm the Egyptians' command of surveying.
- A recent reassessment of Stonehenge (c.2500 BC) suggests that the monument was set out by prehistoric surveyors using peg and rope geometry.
- In the sixth century BC geometric based techniques were used to construct the tunnel of Eupalinos on the island of Samos.

Elements of the Construction Survey

- Survey existing conditions of the future work site, including topography, existing buildings and infrastructure, and underground infrastructure whenever possible (for example, measuring invert elevations and diameters of sewers at manholes);
- Stake out lot corners, stake limit of work and stake location of construction trailer (clear of all excavation and construction)
- Stake out reference points and markers that will guide the construction of new structures
- Verify the location of structures during construction;
- Provide horizontal control on multiple floors.
- Conduct an As-Built survey: a survey conducted at the end of the construction project to verify that the work authorized was completed to the specifications set on plans.

Coordinate Systems Used in Construction

Land surveys and surveys of existing conditions are generally performed according to geodesic coordinates. However for the purposes of construction a more suitable coordinate system will often be used. During construction surveying, the surveyor will often have to convert from geodesic coordinates to the coordinate system used for that project.

Chainage or Station

In the case of roads or other linear infrastructure, a *chainage* (derived from Gunter's Chain - 1 chain is equal to 66 feet or 100 links) will be established, often to correspond with the center line of the road or pipeline. During construction, structures would then be located in terms of *chainage*, *offset* and *elevation*. *Offset* is said to be "left" or "right" relative to someone standing on the *chainage line* who is looking in the direction of increasing *chainage*. Plans would often show *plan* views (viewed from above), *profile* views (a "transparent" section view collapsing all section views of the road parallel to the *chainage*) or *cross-section* views (a "true" section view perpendicular to the *chainage*). In a *plan* view, *chainage* generally increases from left to right, or from the bottom to the top of the plan. *Profiles* are shown with the chainage increasing from left to right, and *cross-sections* are shown as if the viewer is looking in the direction of increasing *chainage* (so that the "left" *offset* is to the *left* and the "right" *offset* is to the *right*).

"Chainage" may also be referred to as "Station".

Building Grids

In the case of buildings, an arbitrary system of grids is often established so as to correspond to the rows of columns and the major load-bearing walls of the building. The grids may be identified alphabetically in one direction, and numerically in the other direction (as in a road map). The grids are usually but not necessarily perpendicular, and are often but not necessarily evenly spaced. Floors and basement levels are also numbered. Structures, equipment or architectural details may be located in reference to the floor and the nearest intersection of the arbitrary axes.

Other Coordinate Systems

In other types of construction projects, arbitrary "north-south" and "east-west" reference lines may be established, that do not necessarily correspond to true coordinates.

Equipment and Techniques Used in Construction Surveying

Surveying equipment, such as levels and theodolites, are used for accurate measurement of angular deviation, horizontal, vertical and slope distances. With computerisation, electronic distance measurement (EDM), total stations, GPS surveying and laser scanning have supplemented (and to a large extent supplanted) the traditional optical instruments.

The builder's level measures neither horizontal nor vertical angles. It simply combines a spirit level and telescope to allow the user to visually establish a line of sight along a level plane. When used together with a graduated staff it can be used to transfer elevations from one location to another. An alternative method to transfer elevation is to use water in a transparent hose as the level of the water in the hose at opposite ends will be at the same elevation. A double right angle prism verifies grid patterns, isolating layout errors.

Equipment and Techniques Used in Mining and Tunneling

Total stations are the primary survey instrument used in mining surveying.

Underground Mining

A total station is used to record the absolute location of the tunnel walls (stopes), ceilings (backs), and floors as the drifts of an underground mine are driven. The recorded data is then downloaded into a CAD program, and compared to the designed layout of the tunnel.

The survey party installs control stations at regular intervals. These are small steel plugs installed in pairs in holes drilled into walls or the back. For wall stations, two plugs are installed in opposite walls, forming a line perpendicular to the drift. For back stations, two plugs are installed in the back, forming a line parallel to the drift.

A set of plugs can be used to locate the total station set up in a drift or tunnel by processing measurements to the plugs by intersection and resection.

Profession

Building Surveying emerged in the 1970s as a profession in the United Kingdom by a group of technically minded General Practice Surveyors. Building Surveying is a recognized profession within Britain and Australia. In Australia in particular, due to risk mitigation/limitation factors the employment of surveyors at all levels of the construction industry is widespread. There are still many countries where it is not widely recognized as a profession. The Services that Building Surveyors undertake are broad but include:

- Construction design and building works
- Project Management and monitoring
- CDM Co-ordinator under the Construction (Design & Management) Regulations 2007
- Property Legislation adviser
- Insurance assessment and claims assistance
- Defect investigation and maintenance adviser
- Building Surveys and measured surveys

- Handling Planning applications
- Building Inspection to ensure compliance with building regulations
- Undertaking pre-acquisition surveys
- Negotiating dilapidations claims

Building Surveyors also advise on many aspects of construction including:

- design
- maintenance
- repair
- refurbishment
- restoration

Clients of a building surveyor can be the public sector, Local Authorities, Government Departments as well as private sector organisations and work closely with architects, planners, homeowners and tenants groups. Building Surveyors may also be called to act as an expert witness. It is usual for building surveyors to undertake an accredited degree qualification before undertaking structured training to become a member of a professional organisation. For Chartered Building Surveyors, these courses are accredited by the Royal Institution of Chartered Surveyors. Other professional organisations that have building surveyor members include CIOB, ABE, HKIS and RICS.

With the enlargement of the European community, the profession of the Chartered Building Surveyor is becoming more widely known in other European states, particularly France. Chartered Building Surveyors, where many English speaking people buy second homes.

Distinction from Land Surveyors

In the United States, Canada, the United Kingdom and most Commonwealth countries land surveying is considered to be a distinct profession. Land surveyors have their own professional associations and licensing requirements. The services of a licensed land surveyor are generally required for boundary surveys (to establish the boundaries of a parcel using its legal description) and subdivision plans (a plot or map based on a survey of a parcel of land, with boundary lines drawn inside the larger parcel to indicated the creation of new boundary lines and roads).

Offshore Geotechnical Engineering

Offshore geotechnical engineering is a sub-field of geotechnical engineering. It is concerned with foundation design, construction, maintenance and decommissioning for human-made structures in the sea. Oil platforms, artificial islands and submarine pipelines are examples of such structures. The seabed has to be able to withstand the weight of these structures and the applied loads.

Platforms offshore Mexico.

Geohazards must also be taken into account. The need for offshore developments stems from a gradual depletion of hydrocarbon reserves onshore or near the coastlines, as new fields are being developed at greater distances offshore and in deeper water, with a corresponding adaptation of the offshore site investigations. Today, there are more than 7,000 offshore platforms operating at a water depth up to and exceeding 2000 m. A typical field development extends over tens of square kilometers, and may comprise several fixed structures, infield flowlines with an export pipeline either to the shoreline or connected to a regional trunkline.

Differences between Onshore and Offshore Geotechnical Engineering

An *offshore* environment has several implications for geotechnical engineering. These include the following:

- Ground improvement (on the seabed) and site investigation are expensive.
- Soil conditions are unusual (*e.g.* presence of carbonates, shallow gas).
- Offshore structures are tall, often extending over 100 metres (330 ft) above their foundation.
- Offshore structures typically have to contend with significant lateral loads (*i.e.* large moment loading relative to the weight of the structure).

- Cyclic loading can be a major design issue.
- Offshore structures are exposed to a wider range of geohazards.
- The codes and technical standards are different from those used for onshore developments.
- Design focuses on ultimate limit state as opposed to deformation.
- Design modifications during construction are either unfeasible or very expensive.
- The design life of these structures often ranges between 25–50 years.
- The environmental and financial costs in case of failure can be higher.

The Offshore Environment

Offshore structures are exposed to various environmental loads: wind, waves, currents and, in cold oceans, sea ice and icebergs. Environmental loads act primarily in the horizontal direction, but also have a vertical component. Some of these loads get transmitted to the foundation (the seabed). Wind, wave and current regimes can be estimated from meteorological and oceanographic data, which are collectively referred to as metocean data. Earthquake-induced loading can also occur – they proceed in the opposite direction: from the foundation to the structure. Depending on location, other geohazards may also be an issue. All of these phenomena may affect the integrity or the serviceability of the structure and its foundation during its operational lifespan – they need to be taken into account in offshore design.

The Nature of the Soil

Following are some to the features characterizing the soil in an offshore environment:

- The soil is made up of sediments, which are generally assumed to be in a saturated state – saline water fills in the pore space.
- Marine sediments are composed of detrital material as well as remains of marine organisms, the latter making up calcareous soils.
- Total sediment thickness varies on a regional scale – it is normally higher near the coastline than it is away from it, where it is also finer grained.
- In places, the seabed can be devoid of sediment, due to strong bottom currents.
- The consolidation state of the soil is either normally consolidated (due to slow sediment deposition), overconsolidated (in places, a relic of glaciation) or underconsolidated (due to high sediment input).

Metocean Aspects

Wave action against an offshore structure.

Wave forces induce motion of floating structures in all six degrees of freedom – they are a major design criterion for offshore structures. When a wave's orbital motion reaches the seabed, it induces sediment transport. This only occurs to a water depth of about 200 metres (660 ft), which is the commonly adopted boundary between *shallow water* and *deep water*.The reason is that the orbital motion only extends to a water depth that is half the wavelength, and the maximum possible wavelength is generally considered to be 400 metres (1,300 ft). In shallow water, waves may generate pore pressure build-up in the soil, which may lead to flow slide, and repeated impact on a platform may cause liquefaction and loss of support.

Currents are a source of horizontal loading for offshore structures. Because of the Bernoulli effect, they may also exert upward or downward forces on structural surfaces and can induce the vibration of wire lines and pipelines. Currents are responsible for eddies around a structure, which cause scouring and erosion of the soil. There are various types of currents: oceanic circulation, geostrophic, tidal, wind-driven, and density currents.

Geohazards

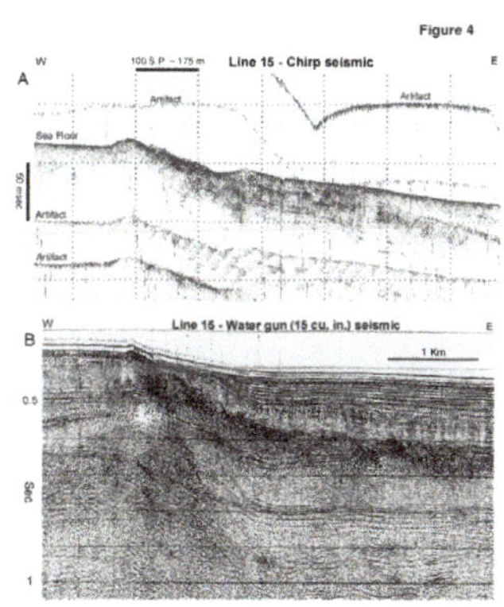

Two types of seismic profiles (top: chirp; bottom: Water gun) of a fault within the seabed in the Gulf of Mexico.

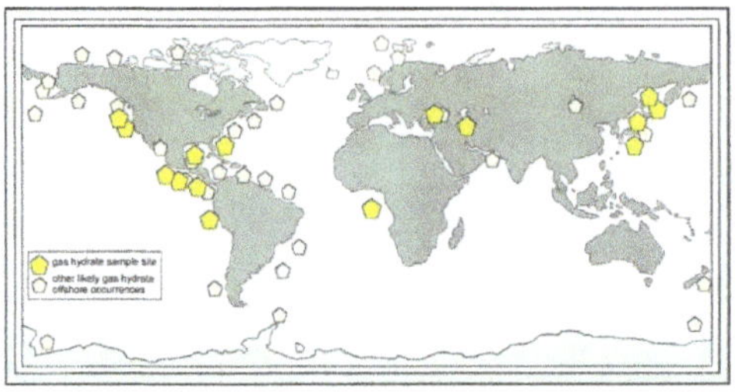

Worldwide distribution of gas hydrates, which are another potential hazard for offshore developments.

An example of a side scan sonar, a device used to survey the seabed.

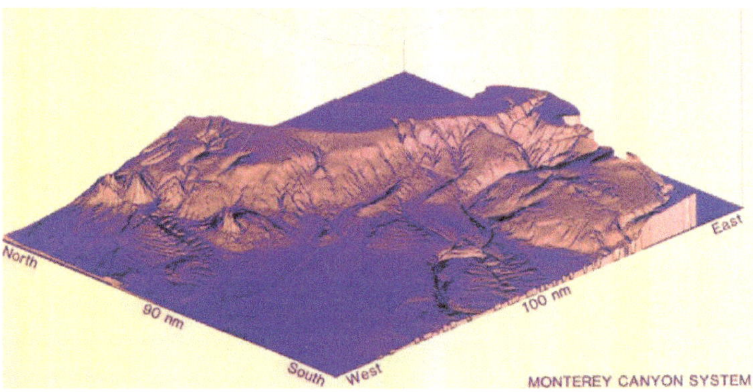

A 3-D image of the Monterey Canyon system, an example of what can be obtained from multibeam echo-sounders.

Geohazards are associated with geological activity, geotechnical features and environmental conditions. Shallow geohazards are those occurring at less than 400 metres (1,300 ft) below the seafloor. Information on the potential risks associated with these phenomena is acquired through studies of the geomorphology, geological setting and tectonic framework in the area of interest, as well as with geophysical and geotechnical surveys of the seafloor. Examples of potential threats include tsunamis, landslides, active faults, mud diapirs and the nature of the soil layering (presence of karst, gas hydrates, carbonates). In cold regions, gouging ice features are a threat to subsea installations, such as pipelines. The risks associated with a particular type of geohazard is a function of how exposed the structure is to the event, how severe this event is and how often it occurs (for episodic events). Any threat has to be monitored, and mitigated for or removed.

Site Investigation

Offshore site investigations are not unlike those conducted onshore. They may be divided into three phases:

- A *desk study*, which includes data compilation.
- *Geophysical surveys*, either shallow and deep seabed penetration.
- *Geotechnical surveys*, which includes sampling/drilling and in situ testing.

Desk Study

In this phase, which may take place over a period of several months (depending on project size), information is gathered from various sources, including reports, scientific literature (journal articles, conference proceedings) and databases, with the purpose of evaluating risks, assessing design options and planning the subsequent phases. Bathymetry, regional geology, potential geohazards, seabed obstacles and metocean data are some of the information that are sought after during that phase.

Geophysical Surveys

Geophysical surveys can be used for various purposes. One is to study the bathymetry in the location of interest and to produce an image of the seafloor (irregularities, objects on the seabed, lateral variability, ice gouges, ...). Seismic refraction surveys can be done to obtain information on shallow seabed stratigraphy – it can also be used to locate material such as sand and gravel for use in the construction of artificial islands. Geophysical surveys are conducted from a research vessel equipped with sonar devices and related equipment, such as single-beam and multibeam echosounders, side-scan sonars, 'towfish' and remotely operated vehicles (ROVs). For the sub-bottom stratigraphy, the tools used include boomers, sparkers, pingers and chirp. Geophysical surveys are normally required before conducting the geotechnical surveys; in larger projects, these phases may be interwoven.

Geotechnical Surveys

Geotechnical surveys involve a combination of sampling, drilling, in situ testing as well as laboratory soil testing that is conducted offshore and/or onshore. They serve to ground truth the results of the geophysical investigations; they also provide a detailed account of the seabed stratigraphy and soil engineering properties. Depending on water depth and metocean conditions, geotechnical surveys may be conducted from a dedicated geotechnical drillship, a semi-submersible, a jackup rig, a large hovercraft or other means. They are done at a series of specific locations, while the vessel maintains a constant position. Dynamic positioning and mooring with four-point anchoring systems are used for that purpose.

Shallow penetration geotechnical surveys may include soil sampling of the seabed surface or in situ mechanical testing. They are used to generate information on the physical and mechanical properties of the seabed. They extend to the first few meters below the mudline. Surveys done to these depths, which may be conducted at the same time as the shallow geophysical survey, may suffice if the structure to be deployed at that location is relatively light. These surveys are also useful for planning subsea pipeline routes.

The purpose of deep penetration geotechnical surveys is to collect information on the seabed stratigraphy to depths extending up to a few 100 meters below the mudline. These surveys are done when larger structures are planned at these locations. Deep drill holes require a few days during which the drilling unit has to remain exactly in the same position.

Sampling and Drilling

Box corer for extracting soil samples from the seabed.

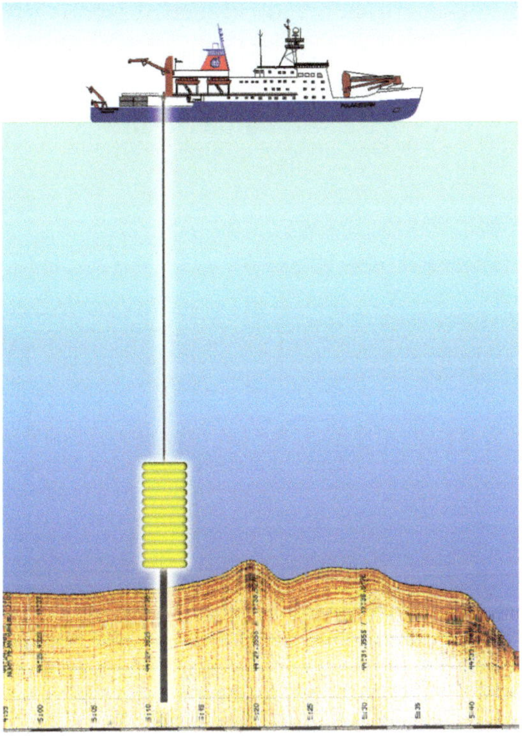

A gravity-driven soil sampler, used for coring the seabed.

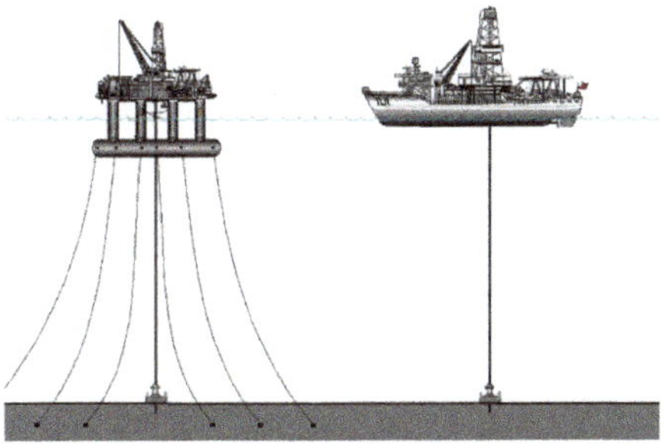

Two types of drilling systems: a semi-submersible (left) and a drillship (right).

Seabed surface sampling can be done with a grab sampler and with a box corer. The latter provides undisturbed specimens, on which testing can be conducted, for instance, to determine the soil's relative density, water content and mechanical properties. Sampling can also be achieved with a tube corer, either gravity-driven, or that can be pushed into the seabed by a piston or by means of a vibration system (a device called a vibrocorer).

Drilling is another means of sampling the seabed. It is used to obtain a record of the seabed stratigraphy or the rock formations below it. The set-up used to sample an offshore structure's foundation is similar to that used by the oil industry to reach and delineate hydrocarbon reservoirs, with some differences in the types of testing. The drill string consists of a series of pipe segments 5 inches (13 cm) in diameter screwed end to end, with a drillbit assembly at the bottom. As the dragbit (teeth extending downward from the drillbit) cut into the soil, soil cuttings are produced. Viscous drilling mud flowing down the drillpipe collects these cuttings and carry them up outside the drillpipe. As is the case for onshore geotechnical surveys, different tools can be used for sampling the soil from a drill hole, notably "Shelby tubes", "piston samplers" and "split spoon samplers".

In Situ Soil Testing

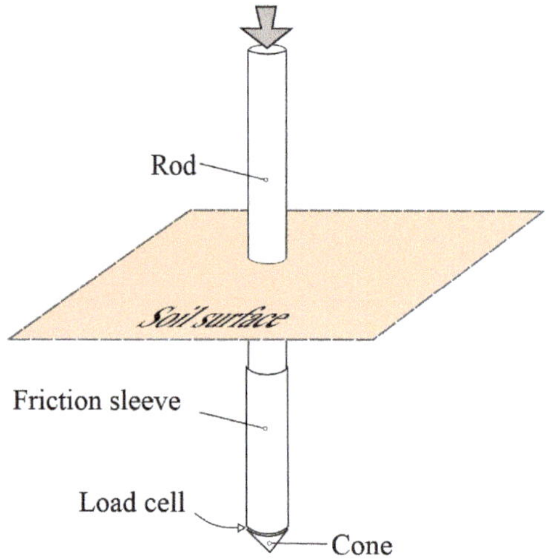

Diagram showing the principle of a cone penetrometer to obtain the soil's strength profile.

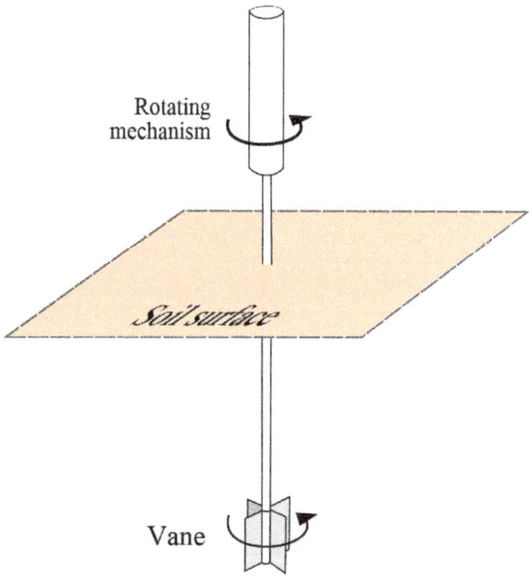

Diagram showing the principle of a shear vane to measure the soil's peak strength and residual strength.

Information on the mechanical strength of the soil can be obtained in situ (from the seabed itself as opposed to in a laboratory from a soil sample). The advantage of this approach is that the data are obtained from soil that has not suffered any disturbance as a result of its relocation. Two of the most commonly used instruments used for that purpose are the cone penetrometer (CPT) and the shear vane.

The CPT is a rod-shaped tool whose end has the shape of a cone with a known apex angle (*e.g.* 60 degrees). As it is pushed into the soil, the resistance to penetration is measured, thereby providing an indication of soil strength. A sleeve behind the cone allows the independent determination of the frictional resistance. Some cones are also able to measure pore water pressure. The shear vane test is used to determine the undrained shear strength of soft to medium cohesive soils. This instrument usually consists of four plates welded at 90 degrees from each other at the end of a rod. The rod is then inserted into the soil and a torque is applied to it so as to achieve a constant rotation rate. The torque resistance is measured and an equation is then used to determine the undrained shear strength (and the residual strength), which takes into account the vane's size and geometry.

Offshore Structures and Geotechnical Considerations

Offshore structures are mainly represented by platforms, notably jackup rigs, steel jacket structures and gravity-based structures. The nature of the seabed has to be taken into account when planning these developments. For instance, a gravity-based structure typically has a very large footprint and is relatively buoyant (because it encloses a large open volume). Under these circumstances, vertical loading of the foundation may not be as significant as the horizontal loads exerted by wave actions and transferred to the seabed. In that scenario, sliding could be the dominant mode of failure. A more specific example is that of the Woodside "North Rankin A" steel jacket structure offshore Australia. The shaft capacity for the piles making up each of the structure's legs was estimated on the basis of conventional design methods, notably when driven into siliceous sands. But the soil at that site was a lower capacity calcareous sand. Costly remediation measures were required to correct this oversight.

Proper seabed characterization is also required for mooring systems. For instance, the design and installation of suction piles has to take into account the soil properties, notably its undrained shear strength. The same is true for the installation and capacity assessment of *plate anchors*.

Submarine Pipelines

Submarine pipelines are another common type of man-made structure in the offshore environment. These structures either rest on the seabed, or are placed inside a trench to protect them from fishing trawlers, dragging anchors or fatigue due current-induced oscillations. Trenching is also used to protect pipelines from gouging by ice keels. In both cases, planning of the pipeline involves geotechnical considerations. Pipelines resting on the seabed require geotechnical data along the proposed pipeline route to evaluate potential stability issues, such as passive failure of the soil below it (the pipeline drops) due to insufficient bearing capacity, or sliding failure (the pipeline shift sideways), due to low sliding resistance. The process of trenching, when required, needs to take into account soil properties and how they would affect ploughing duration. Buckling potential induced by the axial and transverse response of the buried pipeline during its operational lifespan need to be assessed at the planning phase, and this will depend on the resistance of the enclosing soil.

Dry Dock

U.S. Navy submarine USS *Greeneville* in dry dock following collision with the *Ehime Maru*.

A US Navy Littoral Combat Ship in drydock, NASSCO 2012

A dry dock (sometimes dry-dock or drydock) is a narrow basin or vessel that can be flooded to allow a load to be floated in, then drained to allow that load to come to rest on a dry platform. Dry docks are used for the construction, maintenance, and repair of ships, boats, and other watercraft.

History

China

The use of dry docks in China goes at least as far back the 10th century A.D. In 1088, Song Dynasty scientist and statesman Shen Kuo (1031–1095) wrote in his *Dream Pool Essays*:

At the beginning of the dynasty (c. +965) the two Che provinces (now Chekiang and southern Chiangsu) presented (to the throne) two dragon ships each more than 200 ft. in length. The upper works included several decks with palatial cabins and saloons, containing thrones and couches all ready for imperial tours of inspection. After many years, their hulls decayed and needed repairs, but the work was impossible as long as they were afloat. So in the Hsi-Ning reign period (+1068 to +1077) a palace official Huang Huai-Hsin suggested a plan. A large basin was excavated at the north end of the Chin-ming Lake capable of containing the dragon ships, and in it heavy crosswise beams were laid down upon a foundation of pillars. Then (a breach was made) so that the basin quickly filled with water, after which the ships were towed in above the beams. The (breach now being closed) the water was pumped out by wheels so that the ships rested quite in the air. When the repairs were complete, the water was let in again, so that the ships were afloat once more (and could leave the dock). Finally the beams and pillars were taken away, and the whole basin covered over with a great roof so as to form a hangar in which the ships could be protected from the elements and avoid the damage caused by undue exposure. (Wade-Giles spelling.)

Renaissance Europe

Floating Dock. Woodcut from Venice (1560).

The first English and oldest surviving dry dock still in use was commissioned by Henry VII of England at HMNB Portsmouth in 1495 (see Tudor navy). This dry dock currently holds the world's oldest commissioned warship, HMS *Victory*.

Possibly the earliest description of a floating dock comes from a small Italian book printed in Venice in 1560, called *Descrittione dell'artifitiosa machina*. In the booklet, an unknown author asks for the privilege of using a new method for the salvaging of a grounded ship and then proceeds to describe and illustrate his approach. The included woodcut shows a ship flanked by two large floating trestles, forming a roof above the vessel. The ship is pulled in an upright position by a number of ropes attached to the superstructure.

Modern Times

The Alfredo da Silva Dry Dock, of the Lisnave Dockyards in Almada, Portugal, was the largest in the world until 2000, when it was closed after the moving of Lisnave operations to Setúbal.

Harland and Wolff Heavy Industries in Belfast, Northern Ireland, is the site of a large dry dock 556 by 93 metres (1,824 ft × 305 ft). The massive cranes are named after the Biblical figures Samson and Goliath. Goliath stands 96m tall, while Samson is taller at 106m.

The Saint-Nazaire's Chantiers de l'Atlantique owns one of the biggest in the world: 1,200 by 60 metres (3,940 ft × 200 ft). The largest graving dock of the Mediterranean as of 2009 is at the Hellenic Shipyards S.A. (HSY S.A., Athens, Greece). The largest roofed dry dock is at the German Meyer Werft Shipyard in Papenburg, Germany, it is 504 m long, 125 m wide and stands 75 m tall.

Dry Dock 12 at Newport News Shipbuilding is 662 by 76 metres (2,172 ft × 249 ft) the largest dry dock in the USA. The largest floating-dock in North America named The Vigorous. It is operated Vigor Industries in Portland, OR, in the Swan Island industrial area along the Willamette River.

Types

The Stockholm brig "Tre Kronor" in one of the historical dry docks on the island Beckholmen in central Stockholm.

Graving

The classic form of dry dock, properly known as graving dock, is a narrow basin, usually made of earthen berms and concrete, closed by gates or by a caisson, into which a vessel may be floated and the water pumped out, leaving the vessel supported on blocks.

The keel blocks as well as the bilge block are placed on the floor of the dock in accordance with the "docking plan" of the ship. Routine use of dry docks is for the "graving" i.e. the cleaning, removal of barnacles and rust, and re-painting of ships' hulls.

Some fine-tuning of the ship's position can be done by divers while there is still some water left to manoeuvre it about. It is extremely important that supporting blocks conform to the structural members so that the ship is not damaged when its weight is supported by the blocks. Some anti-submarine warfare warships have protruding sonar domes, requiring that the hull of the ship be supported several metres from the bottom of the drydock.

Once the remainder of the water is pumped out, the ship can be freely inspected or serviced. When work on the ship is finished, water is allowed to re-enter the dry dock and the ship is carefully refloated.

U.S. Navy ballistic missile submarine USS *Michigan* inside a flooded dry dock.

Modern graving docks are box-shaped, to accommodate the newer, boxier ship designs, whereas old dry docks are often shaped like the ships that are planned to be docked there. This shaping was advantageous because such a dock was easier to build, it was easier to side-support the ships, and less water had to be pumped away.

Dry docks used for building Navy vessels may occasionally be built with a roof. This is done to prevent spy satellites from taking pictures of the dry dock and any ships or submarines that may be in it. During World War II, fortified dry docks were used by the Germans to protect their submarines from Allied air raids; however, their effectiveness in that role diminished towards the end of the war as bombs became available that could penetrate them.

Today, covered dry docks are usually used only when servicing or repairing a fleet ballistic missile submarine. Another advantage of covered dry docks is that work can take place independently of the weather; this is frequently used by modern shipyards for construction especially of complex, high-value vessels like cruise ships where delays would incur a high cost.

Floating

Floating docks, Gdynia, Poland.

A floating dry dock is a type of pontoon for dry docking ships, possessing floodable buoyancy chambers and a "U"-shaped cross-section. The walls are used to give the dry dock stability when the floor or deck is below the surface of the water. When valves are opened, the chambers fill with water, causing the dry dock to float lower in the water. The deck becomes submerged and this allows a ship to be moved into position inside. When the water is pumped out of the chambers, the dry dock rises and the ship is lifted out of the water on the rising deck, allowing work to proceed on the ship's hull.

The Royal Naval Dockyard, Bermuda as seen by Anna Brassey in 1883, with its floating drydock in the background

A typical floating dry dock involves multiple rectangular sections. These sections can be combined to handle ships of various lengths, and the sections themselves can come in different dimensions. Each section contains its own equipment for emptying the ballast and to provide the required services, and the addition of a bow section can facilitate the towing of the dry dock once assembled. For smaller boats, one-piece floating dry docks can be constructed, potentially coming with their own bow and steering mechanism.

Shipyards operate floating dry docks as one method for hauling or docking vessels. Floating drydocks are important in locations where porous ground prevents the use of conventional drydocks, such as at the Royal Naval Dockyard on the limestone archipelago of Bermuda. Another advantage of floating dry docks is that they can be moved to wherever they are needed and can also be sold second-hand. During World War II, the U.S. Navy used such (floating) dry docks extensively to provide maintenance in remote locations. One of these, the 850-foot AFDB-3, an Advance Base Sectional Dock, saw action in Guam, was mothballed near Norfolk, Virginia, and was eventually towed to Portland, Maine, to become part of Bath Iron Works' repair facilities.

The towboat Dolphin I in a floating dry dock on the Mississippi River in Algiers, New Orleans, Louisiana.

The "Hughes Mining Barge", or HMB-1, is a covered, floating drydock that is also submersible to support the secret transfer of a mechanical lifting device underneath the *Glomar Explorer* ship, as well as the development of the *Sea Shadow* stealth ship.

Blohm + Voss Dock 10, at the Port of Hamburg.

Floating dry dock located in Sevastopol.

A floating dry dock (or boat lift) in a private marina, used to keep small boats out of the water while not in use.

Stern of a cargo ship in the floating dry dock of Bremer Vulkan shipyard, on keel blocks overlooking Oertz rudder and propeller

Alternative Dry Dock Systems

Apart from graving docks and floating dry docks, ships can also be dry docked and launched by:

- Marine railway – For repair of larger ships up to about 3000 tons ship weight
- Shiplift – For repair as well as for new-building. From 800 to 25000 ton ship-weight
- Slipway, patent slip – For repair of smaller boats and the new-building launch of larger vessels

Uses Other than for Ships

Some dry docks are used during the construction of bridges, dams, and other large objects. For example, the dry dock on the artificial island of Neeltje-Jans was used for the construction of the Oosterscheldekering, a large dam in the Netherlands that consists of 65 concrete pillars weighing 18,000 tonnes each. The pillars were constructed in a drydock and towed to their final place on the seabed.

A dry dock may also be used for the prefabrication of the elements of an immersed tube tunnel, before they are floated into position, as was done with Boston's Silver Line.

Coastal Management

Oosterscheldekering sea wall, the Netherlands.

Coastal management is defense against flooding and erosion, and techniques that allow erosion to claim lands.

Historical Background

Coastal engineering, as it relates to harbours, starts with the development of ancient civilizations together with the origin of maritime traffic, perhaps before 3500 B.C. Docks, breakwaters, and other harbour works were built by hand and often in a grand scale. Basic source of modern literature on coastal engineering is the "European Code of Conduct for Coastal Zones" issued by the European Council in 1999. This document was prepared by the Group of Specialists on Coastal Protection and should be used 'as a source of inspiration for national legislation and practice' by decision makers.

The Group of Specialists set up in 1995, pursuant to a decision by the Committee of Ministers of the Council of Europe, who met for the first time on 6 and 7 June 1996. It noted that a great deal of technical and scientific research had been carried out in the field of coastal protection and that various principles and legal texts had been drawn up. It also noted that all of the work undertaken highlighted the need for integrated management and planning of coastal areas, but that, despite all the efforts already made, the situation of coastal areas continued to deteriorate. The Group had acknowledged that this was due to difficulties in implementing the concept of "integrated management", and that it was becoming necessary to provide instruments which would make it easier to apply the principles of integrated coastal management and planning, which had to be pursued to ensure sustainable management of coastal areas. The Group therefore proposed that the Council of Europe, in close co-operation with the Coastal & Marine Union (EUCC) and United Nations Environment Programme (UNEP). The final version of the Code as well of the a MODEL LAW to be used as a guide for modefying local and national legislation, can be free downloaded from the web.

Ancient harbour works are still visible in a few of the harbours that exist today, while others have recently been explored by underwater archaeologists. Most of the grander ancient harbor works have disappeared following the fall of the Western Roman Empire.

Most coastal efforts were directed to port structures, with the exception of a few places where life depended on coastline protection. Venice and its lagoon is one such case. Protection of the shore in Italy, England and the Netherlands can be traced back at least to the 6th century. The ancients understood such phenomena as the Mediterranean currents and wind patterns and the wind-wave cause-effect link.

The Romans introduced many revolutionary innovations in harbor design. They learned to build walls underwater and managed to construct solid breakwaters to protect fully exposed harbors. In some cases wave reflection may have been used to prevent silting. They also used low, water-surface breakwaters to trip the waves before they reached the main breakwater. They became the first dredgers in the Netherlands to maintain the harbour at Velsen. Silting problems here were solved when the previously sealed solid piers were replaced with new "open"-piled jetties.

Middle Ages

The threat of attack from the sea caused many coastal towns and their harbours to be abandoned. Other harbours were lost due to natural causes such as rapid silting, shoreline advance or retreat, etc. The Venetian Lagoon was one of the few populated coastal areas with continuous prosperity and development where written reports document the evolution of coastal protection works.

Engineering and scientific skills remained alive in the east, in Byzantium, where the Eastern Roman Empire survived for six hundred years while Western Rome decayed.

Modern Age

Although great strides were made in the general scientific arena, little improvement was done beyond the Roman approach to harbour construction after the Renaissance. In the early 19th century, the advent of the steam engine, the search for new lands and trade routes, the expansion of the British Empire through her colonies, and other influences, all contributed to the revitalization of sea trade and a renewed interest in port works.

Twentieth Century

Evolution of shore protection and the shift from structures to beach nourishment. Prior to the 1950s, the general practice was to use hard structures to protect against beach erosion or storm damages. These structures were usually coastal armoring such as seawalls and revetments or sand-trapping structures such as groynes. During the 1920s and '30s, private or local community interests protected many areas of the shore using these techniques in a rather ad hoc manner. In certain resort areas, structures had proliferated to such an extent that the protection actually impeded the recreational use of the beaches. Erosion of the sand continued, but the fixed back-beach line remained, resulting in a loss of beach area.

The obtrusiveness and cost of these structures led in the late 1940s and early 1950s, to move toward a new, more dynamic, method. Projects no longer relied solely on hard coastal defence structures, as techniques were developed which replicated the protective characteristics of natural beach and dune systems. The resultant use of artificial beaches and stabilized dunes as an engineering approach was an economically viable and more environmentally friendly means for dissipating wave energy and protecting coastal developments.

Over the past hundred years the limited knowledge of coastal sediment transport processes at the local authorities level has often resulted in inappropriate measures of coastal erosion mitigation. In many cases, measures may have solved coastal erosion locally but have exacerbated coastal erosion problems at other locations-up to tens of kilometers away- or have generated other environmental problems.

Current Challenges in Coastal Management

The coastal zone is a region where interaction of the sea and land processes occurs. They occupy less than 15% of the Earth's land surface; yet accommodate more than 40% of the world population. Nearly 1.2 billion people live within 100 km of a shoreline and 100 m of sea level with an average density nearly 3 times higher than the global average for population. With three-quarters of the world population expected to reside in the coastal zone by 2025, human activities originating from this small land area will impose an inordinate amount of pressures on the global system. Coastal zones contain rich resources to produce goods and services and are home to most commercial and industrial activities. In the European Union, almost half of the population now lives within 50 kilometres of the sea and coastal zone resources produce much of the Union's economic wealth. The fishing, shipping and tourism industries all compete for vital space along Europe's estimated 89 000 kilometres of coastline, and coastal zones contain some of Europe's most fragile and valuable natural habitats. Shore protection consists up to the 50's of interposing a static structure between the sea and the land to prevent erosion and or flooding, and it has a long history. From that period new technical or friendly policies have been developed to preserve the environment when possible. It is already important where there are extensive low-lying areas that require protection. For instance: Venice, New Orleans, Nagara river in Japan, the Netherlands, Caspian Sea. Protection against sea level rise in the 21st century will be especially important, as sea level rise is currently accelerating. This will be a challenge to coastal management, since seawalls and breakwaters are generally expensive to construct, and the costs to build protection in the face of sea-level rise would be enormous.

Changes in sea level have a direct adaptative response from beaches and coastal systems, as we can see in the succession of a lowering sea level. When the sea level rises, coastal sediments are in part pushed up by wave and tide energy, so sea-level rise processes have a component of sediment transport landwards. This results in a dynamic model of rise effects with a continuous sediment displacement that is not compatible with static models where coastline change is only based on topographic data.

Planning Approaches

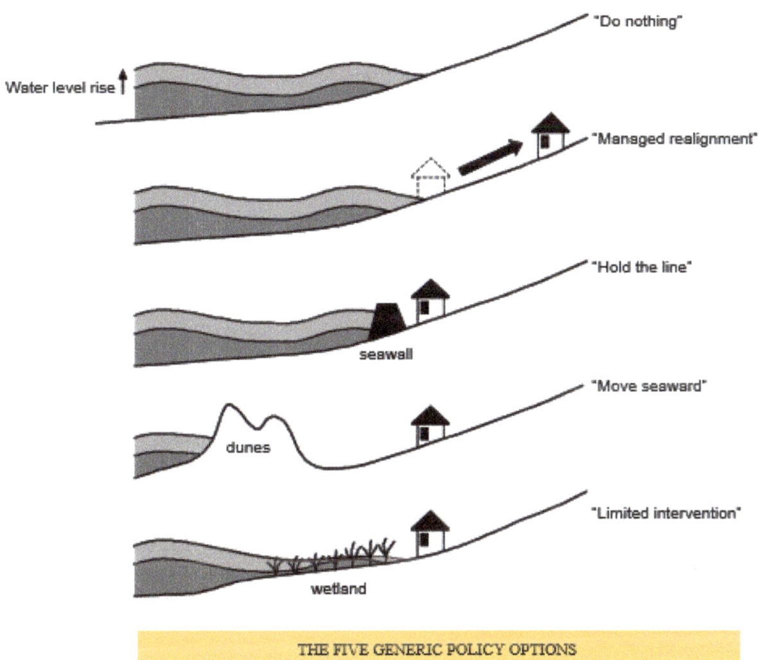

Five general coastal management strategies

There are five generic strategies for coastal defense:

- inaction leading to eventual abandonment

- Managed retreat or realignment, which plans for retreat and adopts engineering solutions that recognise natural processes of adjustment, and identifies a new line of defence where to construct new defences
- Hold the line, shoreline protection, whereby seawalls are constructed around the coastlines
- Move seawards, this happens by constructing new defenses seaward the original ones
- Limited intervention, accommodation, by which adjustments are made to be able to cope with inundation, raising coastal land and buildings vertically

The decision to choose a strategy is site-specific, depending on pattern of relative sea-level change, geomorphological setting, sediment availability and erosion, as well a series of social, economic and political factors.

Alternatively, integrated coastal zone management approaches may be used to prevent development in erosion- or flood-prone areas to begin with. Growth management can be a challenge for coastal local authorities who often struggle to provide the infrastructure required by new residents seeking seachange lifestyles. Sustainable transport investment to reduce the average footprint of coastal visitors is often a good way out of coastal gridlock. Examples include Dongtan and the Gold Coast Oceanway.

The 'Managed Retreat' option, involving no protection, is cheap and expedient. The coast takes care of itself and coastal facilities are abandoned to coastal erosion, with either gradual landward retreat or evacuation and resettlement elsewhere. This is the usual response when land of little value will be lost. The only pollution produced is from the resettlement process. Where endangered property has high value, it is less often applied.

Managed Retreat

Managed retreat is an alternative to constructing or maintaining coastal structures. Managed retreat allows an area to become flooded. This process is usually in low lying estuarine or deltaic areas and floods land that has at some point in the past been reclaimed from the sea. Managed retreat is often a response to a change in sediment budget or to sea level rise. The technique is used when the land adjacent to the sea is low in value. A decision is made to allow the land to erode and flood, creating new shoreline habitats. This process may continue over many years and natural stabilization will occur.

The earliest managed retreat in the UK was an area of 0.8 ha at Northey Island in starfish island flooded in 1991. This was followed by Tollesbury and Orplands in Essex, where the sea walls were breached in 1995. In the Ebro delta (Spain) coastal authorities have planned a managed retreat in response to coastal erosion (MMA 2005, Sitges, Meeting on Coastal Engineering; EUROSION project).

Cost – The main cost is generally the purchase of land to be flooded. Compensation for relocation of residents may be needed. Any other human made structure which will be engulfed by the sea may need to be safely dismantled to prevent sea pollution. In some cases, a retaining wall or bund must be constructed inland in order to protect land beyond the area to be flooded, although such structures can generally be lower than would be needed on the existing coast. Monitoring of the evolution of the flooded area is another cost. Costs may be lowest if existing defenses are left to fail naturally, but often the realignment project will be more actively managed, for example by creating an artificial breach in existing defences to allow the sea in at a particular place in a controlled fashion, or by pre-forming drainage channels for created salt-marsh.

Hold the Line

Human strategies on the coast have been heavily based on a static engineered response, whereas the coast is in, or strives towards, a dynamic equilibrium (Schembri, 2009). Solid coastal structures are built and persist because they protect expensive properties or infrastructures, but they often relocate the problem downdrift or to another part of the coast. Soft options like beach nourishment, while also being temporary and needing regular replenishment, appear more acceptable, and go some way to restore the natural dynamism of the shoreline. However, in many cases there is a legacy of decisions that were made in the past which have given rise to the present threats to coastal infrastructure and which necessitate immediate shore protection. For instance, the seawall and promenade of many coastal cities in Europe represents a highly engineered use of prime seafront space, which might be preferably designated as public open space, parkland and amenities if it were available today. Such open space might also allow greater flexibility in terms of future land-use change, for instance through managed retreat, in the face of threats of erosion or inundation as a result of sea-level rise. Foredunes areas represent a natural reserve which can be called upon in the face of extreme events; building on these areas leaves little option but to undertake costly protective measures when extreme events (whether amplified by gradual global change or not) threaten. Managed retreat can comprise 'setbacks', rolling easements and other planning tools including building within a particular design life. Maintenance of those structures or soft techniques can arrive at a critical point (economically or environmental) to change adopted strategy.

- Structural or hard engineering techniques, i.e. using permanent concrete and rock constructions to "fix" the coastline and protect the assets locate behind. These techniques--seawalls, groynes, detached breakwaters, and revetments—represent a significant share of protected shoreline in Europe (more than 70%).
- Soft engineering techniques (e.g. sand nourishments), building with natural processes and relying on natural elements such as sands, dunes and vegetation to prevent erosive forces from reaching the backshore. These techniques include beach nourishment and sand dune stabilization.

Move Seaward

The futility of trying to predict future scenarios where there is a large human influence is apparent. Even future climate is to a certain extent a function of what humans choose to make of it, for example by restricting greenhouse gas emissions to control climate change. In some cases - where new areas are needed for new economic or ecological development - a move seaward strategy can be adopted. Examples from erosion include: Koge Bay (Dk) Western Scheldt estuary (Nl), Chatelaillon (F), Ebro delta (E)

There is an obvious downside to this strategy. Coastal erosion is already widespread, and there are many coasts where exceptional high tides or storm surges result in encroachment on the shore, impinging on human activity. If the sea rises, many coasts that are developed with infrastructure along or close to the shoreline will be unable to accommodate erosion. They will experience a so-called "coastal squeeze" whereby the ecological or geomorphological zones that would normally retreat landwards encounter solid structures and are squeezed out. Wetlands, salt marshes, mangroves and adjacent fresh water wetlands are particularly likely to suffer from this squeeze.

An upside to the strategy is that moving seaward (and upward) can create land of high value which can bring the investment required to cope with climate change.

Limited Intervention

Limited intervention is an action taken whereby the management only solves the problem to some extent, usually in areas of low economic significance. Measures taken using limited intervention often encourage the succession of haloseres, including salt marshes and sand dunes. This will normally result in the land behind the halosere being more sufficiently protected, as wave energy will be dissipated by the accumulated sediment and additional vegetation residing in the newly formed habitat. Although the new halosere is not strictly man-made, as many natural processes will contribute to the succession of the halosere, anthropogenic factors are partially responsible for the formation as an initial factor was needed to help start the process of succession. This must not be confused with 'accommodate' which is about property for example effective insurance, early warning systems and not about habitat.

Construction Techniques

The following is a catalogue of relevant techniques that could be employed as coastal management techniques. *The costs given are very rough estimates made during 2005, based on UK Pound sterling.*

Hard Engineering Methods

Groynes

Groynes are barriers or walls perpendicular to the sea, often made of greenharts, concrete, rock or wood. Beach material builds up on the downdrift side, where littoral drift is predominantly in one direction, creating a wider and a more plentiful beach, therefore enhancing the protection for the coast because the sand material filters and absorbs the wave energy. However, there is a corresponding loss of beach material on the updrift side, requiring that another groyne to be built there. Moreover, groynes do not protect the beach against storm-driven waves and if placed too close together will create currents, which will carry sand material offshore.

Groyne at Mundesley, Norfolk, UK

Groynes are extremely cost-effective coastal defence measures, requiring little maintenance, and are one of the most common coastal defence structures. However, groynes are increasingly viewed as detrimental to the aesthetics of the coastline, and face strong opposition in many coastal communities.

Many experts consider groynes to be a "soft" solution to coastal erosion because of the enhancement of the existing beach.

But groyne construction creates a problem known as terminal groyne syndrome. The terminal groyne prevents longshore drift from bringing material to other nearby places. This is a common problem along the Hampshire and Sussex coastline in the UK; an example is Worthing. Beach material does not get washed away because of these groynes but can cause damage to other parts of the coast.

Sea Walls

Walls of concrete or rock, built at the base of a cliff or at the back of a beach, are used to protect a settlement against erosion or flooding. They are usually about 3–5 metres (10–16 ft) high. Older-style vertical seawalls reflected all the energy of the waves back out to sea, and for this purpose were often given recurved crest walls which also increase the local turbulence, and thus increasing entrainment of sand and sediment. During storms, sea walls help longshore drift.

Modern seawalls aim to re-direct most of the incident energy, resulting in low reflected waves and much reduced turbulence and thus take the form of sloping revetments. Current designs use porous designs of rock, concrete armour (Seabees, SHEDs, Xblocs) with intermediate flights of steps for beach access, whilst in places where high rates of pedestrian access are required, the steps take over the whole of the frontage, but at a flatter slope if the same crest levels are to be achieved.

Care needs to be taken in the location of a seawall, particularly in relation to the swept prism of the beach profile, the consequences of long-term beach recession and amenity crest level. These factors must be considered in assessing the cost-benefit ratio, which must be favorable in order to justify construction of a seawall.

Sea walls can cause beaches to dissipate, rendering them useless for beach goers. Their presence also scars the very landscape that they are trying to save.

Modern examples can be found at Cronulla (NSW, 1985-6), Blackpool (1986–2001), Lincolnshire (1992–1997) and Wallasey (1983–1993). The sites at Blackpool and Cronulla can be visited both by Google Earth and by local webcams (Cronulla, Cleveleys).

An example is the seawall at Sandwich, Kent, where the Seabee seawall is buried at the back of the beach under the shingle with crest level at road kerb level.

Sea walls are probably the second most traditional method used in coastal management.

Sea walls cost £10,000 per metre (depending on material, height and width), £10,000,000 per km (depending on material, height and width).

Revetments

Wooden slanted or upright blockades, built parallel to the sea on the coast, usually towards the back of the beach to protect the cliff or settlement beyond. The most basic revetments consist of timber slants with a possible rock infill. Waves break against the revetments, which dissipate and absorb the energy. The cliff base is protected by the beach material held behind the barriers, as the revetments trap some of the material. They may be watertight, covering the slope completely, or porous, to allow water to filter through after the wave energy has been dissipated. Most revetments do not significantly interfere with transport of longshore drift. Since the wall greatly absorbs the energy instead of reflecting, it erodes and destroys the revetment structure; therefore, major maintenance will be needed within a moderate time of being built, this will be greatly determined by the material the structure was built with and the quality of the product.

The *Cost* – Confirmed by material used; est. $2340–$4000. Average $10 per meter built – around £6 GBP.

Rock Armour

Also known as riprap, rock armour are large rocks piled or placed at the foot of dunes or cliffs with native stones of the beach. This is generally used in areas prone to erosion to absorb the wave energy and hold beach material. Although effective, this solution is unpopular due to the fact that it is unsightly. Also, longshore drift is not hindered. Rock armour has a limited lifespan, it is not effective in storm conditions, and it reduces the recreational value of a beach. The cost is around £3000 per metre, depending on the type of rocks used.

Gabions

Boulders and rocks are wired into mesh cages and usually placed in front of areas vulnerable to heavy erosion: sometimes at cliffs edges or jag out at a right angle to the beach like a large groyne. When the seawater breaks on the gabion, the water drains through leaving sediments, also the rocks and boulders absorb a moderate amount of the wave energy.

Gabions need to be securely tied to prevent abrasion of wire by rocks, or detachment of plastic

Downsides include wear rates and visual intrusiveness.

Cost – est. £11 per m

Offshore Breakwater

Massive concrete blocks and natural boulders are sunk offshore to alter wave direction and to filter the energy of waves and tides. The waves break further offshore and therefore reduce their erosive power. This leads to wider beaches, which absorb the reduced wave energy, protecting cliff and settlements behind. The Dolos which was invented by a South African engineer in East London has replaced the use of enormous concrete blocks because the dolos is much more resistant to wave action and requires less concrete to produce a superior result. Similar concrete objects like the Dolos are the A-jack, Akmon, Xbloc and the Tetrapod, Accropode. *Cost* – est. £2,000 per m. Water depth may increase the cost.

Cliff Stabilization

Cliff stabilization can be accomplished through drainage of excess rainwater of through terracing, planting, and wiring to hold cliffs in place. Cliff drainage is used to hold a cliff together using plants, fences and terracing, this is used to help prevent landslides and other localized damage.

Entrance Training Walls

Rock or concrete walls built to constrain a river or creek discharging across a sandy coastline. The walls help to stabilise and deepen the channel which benefits navigation, flood management, river erosion and water quality but can cause coastal erosion due to the interruption of longshore drift. One solution is the installation of a sand bypassing system to pump sand under and around the entrance training walls.

Cost – Expensive – Gold Coast Seaway was a A$50M project in the 1980s and the adjacent sand bypassing project costs A$3M per year to pump 500,000 cubic meters of sand across the trained entrance.

Floodgates

Storm surge barriers, or floodgates, were introduced after the North Sea Flood of 1953 and are a prophylactic method to prevent damage from storm surges or any other type of natural disaster that could harm the area they "protect". They are habitually open and allow free passage, but close when the land is under threat of a storm surge. The Thames Barrier is an example of such a structure.

Soft Engineering Methods

Beach Replenishment

Beach replenishment or nourishment is one of the most popular soft engineering techniques of coastal defence management schemes. This involves importing sand off the beach and piling it on top of the existing sand. The imported sand must be of a similar quality to the existing beach material so it can integrate with the natural processes occurring there, without causing any adverse effects. Beach nourishment can be used alongside the groyne schemes. The scheme requires constant maintenance: 1 to 10-year life before first major recharge. *Cost* – est. £50-£2000 per metre, plus control structures, ongoing management and minor works.

Sand Dune Management

Sand dune stabilisation or sand dune management works using a number of different methods in order to prevent the loss of sediment on the beach. Firstly the introduction of public amenities such as car parks, footpaths, Dutch Ladders and boardwalks, stop the removal of sediment by humans. Secondly, education of visitors with noticeboards, leaflets and beach wardens explain to visitors how to avoid damaging the area. Thirdly, by using fences constructed of simple materials such as wood, sand traps can create Blowouts. Furthermore, natural plants such as Ammophila, sometimes known as Marram Grass, is introduced to blowouts in order to bind the sediment together - preventing loss. Finally, areas of the beach can be simply closed to the public to allow rejuvenation to occur.

Cost – est. of £1.1 million per annum

Beach Drainage

Beach drainage or beach face dewatering lowers the water table locally beneath the beach face. This causes accretion of sand above the drainage system.

Grant (1946) – the elevation of the beach watertable had an important bearing on deposition and erosion across the foreshore. A high watertable coincided with periods of accelerated beach erosion, and conversely, a low watertable coincided with pronounced aggradation of the foreshore A lower watertable (unsaturated beach face) facilitates deposition by reducing flow velocities during backwash and prolonging laminar flow. In contrast, a high watertable results in condition favoring beach erosion. With the beach in a saturated state, Grant proposed that backwash velocity is accelerated by the addition of groundwater seepage out of the beach within the effluent zone.

Turner and Leatherman (1997) moving from the origins and development of the dewatering concept to field and laboratory studies available at the time of writing concluded that there was too little evidence for being convinced that the systems had a positive effect. None of the case studies provide full scientific evidence of indisputable positive results regarding beach stabilisation although in some cases an overall positive performance was reported. In many cases no adequate long-term monitoring was undertaken at a frequency high enough to discriminate the response to high energy erosive events.

A useful side effect of the system is that the collected seawater is very pure because of the sand filtration effect. It may be discharged back to sea but can also be used to oxygenate stagnant inland lagoons/marinas or used as feed for heat pumps, desalination plants, land-based aquaculture, aquariums or seawater swimming pools.

Beach drainage systems have been installed in many locations around the world to halt and reverse erosion trends in sand beaches. Twenty four beach drainage systems have been installed since 1981 in Denmark, USA, UK, Japan, Spain, Sweden, France, Italy and Malaysia.

Costs

The costs of installation and operation per meter of shoreline protection will vary due to

- system length (non-linear cost elements)
- pump flow rates (sand permeability, power costs)
- soil conditions (presence of rock or impermeable strata)
- discharge arrangement /filtered seawater utilization
- drainage design, materials selection & installation methods

- geographical considerations (location logistics)
- regional economic considerations (local capabilities /costs)
- study requirements /consent process.

The costs associated with a beach drainage system are generally considerably lower than hard engineered structures. They also compare very favorably with beach nourishment projects, particularly when long-term project economics are considered (nourishment projects often have a limited life or a program of re-nourishment).

Monitoring Coastal Zones

Coastal zone managers are faced with difficult and complex choices about how best to reduce property damage in the shorelines. One of the problems they face is error and uncertainty in the information available to them on the processes that cause erosion of beaches. Video-based monitoring lets collect data continuously at low cost and produce analyses of shoreline processes over a wide range of averaging intervals.

Event Warning Systems

Event warning systems, such as tsunami warnings and storm surge warnings, can be used to minimize the human impact of catastrophic events that cause coastal erosion. Storm surge warnings can also be used to determine when to close floodgates to reduce the physical impact of such events.

Wireless sensor networks can be deployed quickly to set up a coastal erosion monitoring system, and scaled accordingly.

Shoreline Mapping

Defining the shoreline is a difficult task due to the dynamic nature of the coast and the intended application of the shoreline (Graham et al. 2003; Boak & Turner 2005). Given this idea the shoreline must therefore be considered in a temporal sense whereby the scale is dependent on the context of the investigation (Boak & Turner 2005). The following definition of the coast and shoreline is most commonly employed for the purposes of shoreline mapping. The coast comprises the interface between land and sea, and the shoreline is represented by the margin between the two (Woodroffe, 2002). Due to the dynamic nature of the shoreline coastal investigators adopt the use of shoreline indicators to represent the true shoreline position (Boak & Turner 2005).

Shoreline Indicator

The choice of shoreline indicator is a primary consideration in shoreline mapping. According to Leatherman (2003) it is important that indicators are easily identified in the field and on aerial photography. Shoreline indicators may be physical beach morphological features such as the berm crest, scarp edge, vegetation line, dune toe, dune crest and cliff or the bluff crest and toe. Alternatively, non-morphological features may also be used. These indicators are based on water level including the high water line, mean high water line, wet/dry boundary, and the physical water line (Pajak & Leatherman 2000). Figure 1 provides a sketch of the spatial relationship between many of the commonly used shoreline indicators.

The high water line (HWL), defined as the wet/dry line (H in Figure 1) is the most commonly used shoreline indicator because it is visible in the field, and can be interpreted on both colour and grey scale aerial photographs (Leatherman, 2003; Crowell et al. 1991). The HWL represents the landward extent of the most recent high tide and is characterised by a change in sand colour due to repeated, periodic inundation by high tides. The HWL is portrayed on aerial photographs by the most landward change in colour or grey tone (Boak & Turner 2005).

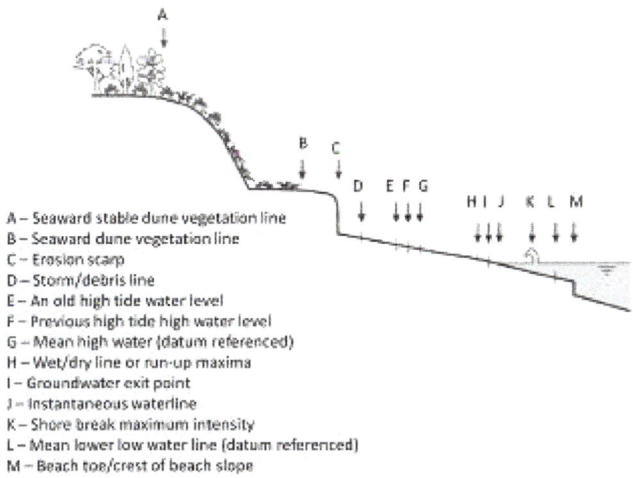

A diagram representing the spatial relationship between many of the commonly used indicators. (Adapted from Boak and Turner 2005)

Importance and Application

The location of the shoreline and its changing position over time is of fundamental importance to coastal scientists, engineers and managers (Boak & Turner 2005; Pajack & Leatherman 2002). Present day shoreline monitoring campaigns provide information about historic shoreline location and movement, and about predictions of future change (Appeaning Addo et al. 2008). More specifically the position of the shoreline in the past, at present and where it is predicted to be in the future is useful for in the design of coastal protection, to calibrate and verify numerical models to assess sea level rise, map hazard zones and formulate policies to regulate coastal development. Accurate and consistent delineation of the shoreline is integral to all of these tasks. The location of the shoreline also provides information regarding shoreline reorientation adjacent to structures, beach width, volume and rates of historical change (Boak & Turner 2005; Pajack & Leatherman 2002).

Data Sources

A variety of data sources are available for examining shoreline position however, the availability of historical data is limited at many coastal sites and so the choice of data source is largely limited to what is available for the site at a given time (Boak & Turner 2005). Shoreline mapping techniques applied to data sources have moved towards automation in association with technological advances and the need to reduce uncertainty. Although these changes have resulted in improvement in coastal data processing and storage capabilities, the frequent change in technology has prevented the emergence of one standard method of shoreline mapping. This has occurred because each data source and associated method have their own unique capabilities and shortcomings (Moore 2000). A number of the data sources used for shoreline mapping and their associated advantages and disadvantages are discussed below.

Historical Maps

In the event that a study requires the shoreline position to be mapped before the development of aerial photographs, or if the location has poor photograph coverage it is necessary to employ historical maps in order to detail shoreline position (Moore 2000). The main advantage and reason for using historical maps is that they are able to provide a historic record that is not available from other data sources. Many potential errors however are associated with historical coastal maps and charts. Such errors may be associated with scale, datum changes, distortions from uneven shrinkage, stretching, creases, tears and folds, different surveying standards, different publication standards, and projection errors (Boak & Turner 2005). The severity of these errors depends on the accuracy standards met by each map and the physical changes that have occurred since the publication of the map (Anders & Byrnes 1991). The oldest reliable source of shoreline data in the United States dates back to the early-to-mid-19th century and is the U.S Coast and Geodetic Survey/National Ocean Service T-sheets (Morton 1991). In the United Kingdom, many maps and charts were deemed to be inaccurate until around 1750. The founding of the Ordnance Survey in 1791 has since improved the accuracy of the mapping.

Aerial Photographs

Aerial photographs have been used since the 1920s to provide topographical information about an area. They are therefore a good database for compilation of historical shoreline change maps. Aerial photographs are the most commonly used data source in shoreline mapping because many coastal areas have extensive aerial photo coverage therefore providing a valuable record of shoreline position (Moore 2000). In general, aerial photographs provide good spatial coverage of the coast however temporal coverage is very much site specific depending on the flight path of the aeroplane. A second disadvantage associated with aerial photography is that the interpretation of the shoreline position is subjective given the dynamic nature of the coastal environment. This combined with various distortions inherent in aerial photographs can lead to significant error levels (Moore 2000). The minimisation of further errors is discussed below.

Object Space Displacements

Conditions outside of the camera can cause objects in an image to be displaced from their true ground position. Such conditions may include ground relief, camera tilt and atmospheric refraction.

Relief displacement is prominent when photographing a variety of elevations. This situation causes objects above ground level to be displaced outward from the centre of the photograph and objects below ground level to be displaced toward the centre of the image (Figure 2). The severity of the displacement is affected negatively with decreases in flight altitude and as radial distance from the centre of the photograph increases. This distortion can be minimised by photographing numerous swaths and creating a mosaic of the images. This technique will create a focus for the centre of each photograph where distortion is minimised. It is important to note that this error is not common in shoreline mapping is the relief is fairly constant. It is however important to consider when mapping cliffs (Moore 2000).

An example of relief displacement. All objects above ground level are displaced outwards from the centre of the photograph. The displacement becomes more evident near the edges.

Ideally aerial photographs are taken so the optical axis of the camera is perfectly perpendicular to the ground surface thereby creating a vertical photograph. Unfortunately this is not often the case and virtually all aerial photographs experience tilt whereby up to 3° is not uncommon (Camfield et al. 1996). In this situation the scale of the image will be larger on the upward side of the tilt axis and smaller on the downward side. Moore, (2000) notes that many coastal researchers have not realised the severity of this error and therefore do not consider it in their methods.

Radial Lens Distortion

Lens distortion varies as a function of radial distance from the iso-centre of the photograph meaning that the centre of the image is relatively distortion free, but as the angle of view increases the distortion becomes more prominent. This is a significant source of error in earlier aerial photography but as technology has increased and camera lens have become more refined it has become less of an issue with later photographs. Such a distortion is impossible to correct for without knowing the make and model of the lens used to capture the image. However, if overlapping images have been acquired one can digitize the centre portions of the aerial photographs (Crowell et al. 1991).

Delineation of the Shoreline

The dynamic nature of the coast has meant that accurate mapping of an instantaneous shoreline position has been associated with significant uncertainty. This uncertainty arises because at any given time the position of the shoreline is influenced by the short-term effect of the tide and a wide variety of long-term effects such as relative sea-level rise and along shore littoral sediment movement. Not only does this affect the accuracy of computed historic shoreline position but also any predicted future positions (Appeaning Addo et al. 2008). As mentioned earlier the HWL is most commonly used as a shoreline indicator. This can usually be seen as a significant tonal change on aerial photographs. There are however many errors associated with using the wet/dry line as a proxy for the HWL and shoreline. The errors of largest concern are the short-term migration of the wet/dry line, interpretation of the wet/dry line on a photograph and measurement of the interpreted line position (Leatherman 2003; Moore 2000). Systematic errors such as the migration of the wet/dry line may arise from tidal and seasonal changes. Storm-induced erosion is another factor which may cause the wet/dry line to migrate landward. Field investigations have shown that these changes can be minimised by using only summertime data (Moore 2000; Leatherman 2003). Furthermore, the error bar can be significantly reduced by using the longest record of reliable data to calculate erosion rates (Leatherman 2003). Finally it is important to note that errors may arise due to the difficulty of measuring a single line on a photograph. For example, where the pen line is 0.13 mm thick this translates to an error of ±2.6 m on a 1:20000 scale photograph.

Beach Profiling Surveys

Beach profiling surveys are typically repeated at regular intervals along the coast in order to measure short-term (daily to annual) variations in shoreline position and beach volume. (Smith & Zarillo 1990). Beach profiling is a very accurate source of information however measurements are generally subject to the limitations of conventional surveying techniques. Shoreline data derived from beach profiling is often spatially and temporally limited due to the high cost associated with such a labour-intensive activity. Shorelines are generally derived by interpolating between a series of discrete beach profiles. It is important to note however that the distance between the profiles is usually quite large and so the accuracy of the interpolating becomes compromised. In contrast to aerial photographs, survey data is limited to smaller lengths of shoreline generally less than ten kilometres (Boak & Turner 2005). Beach profiling data is commonly available in from regional councils in New Zealand such as those compiled by the Hawkes Bay Regional Council.

Remote Sensing

Technological advancement over the last decade has led to the development of a range of airborne, satellite and land based remote sensing techniques (Smith & Zarillo 1990). Some of the remotely sensed data sources are listed below:

- Multispectral and hyperspectral imaging
- Microwave sensors
- Global positioning system (GPS)
- Airborne light detection and ranging technology (LIDAR)

Remote sensing techniques are attractive as they are cost effective, reduce manual error and remove the subjective approach of conventional field techniques (Maiti et al. 2009). Remote sensing is a relatively new concept and so extensive historical observations are unavailable. Given this idea, it is important that coastal morphology observations are quantified by coupling remotely sensed data with other sources of information detailing historic shoreline position from archived sources (Appeaning Addo et al. 2008).

Video Analysis

Video analysis provides quantitative, cost-effective, continuous and long-term monitoring beaches (Turner et al. 2004). The advancement of coastal video systems over the past 15 years has resulted in the extraction of large amounts of geophysical data from images. Such data includes that about coastal morphology, surface currents and wave parameters. The main advantage of video analysis lies in the ability to reliably quantify these parameters with high resolution and coverage in both space and time. This in particular highlights their potential importance as an effective coastal monitoring system and an aid to coastal zone management (Van Koningsveld et al. 2007). Interesting case studies have been carried out using video analysis. Turner et al. (2004) used a video-based ARGUS coastal imaging system to monitor and quantify the regional-scale coastal response to sand nourishment and construction of the world-first Gold Coast artificial (surfing) reef in Australia. In addition, Smit et al. (2007) demonstrated the added value of high resolution video observations for making short-term predictions of near shore hydrodynamic and morphological processes, at temporal scales of meters to kilometres and days to seasons.

Recent advances in video analysis give coastal zone managers the opportunity to obtain bathymetries from video analysis (Plant et al. 2008, Holman et al. 2013 and Bergsma et al. 2016). High spatial and temporal resolution as well as the versatile character of the video systems make it a valuable tool to obtain inter-tidal topographies and sub-tidal bathymetries and measure coastal zone resilience [as in available beach volume as well as sub-tidal bar configuration]. The video-based depth estimations have been applied in micro/meso tidal environments at DUCK, NC (Holman et al. 2013) and highly energetic wave climates with a macro tidal regime at Porthtowan in the United Kingdom (Bergsma et al. 2016). Bergsma (2016) shows the application of video-based depth estimations during extreme storms in a test-case for the 2013-2014 extreme storms that hit Western Europe (Masselink et al. 2015, Castelle et al. 2015).

Transportation Engineering

The engineering of this roundabout in Bristol, England, attempts to make traffic flow free-moving

Transportation engineering or transport engineering is the application of technology and scientific principles to the planning, functional design, operation and management of facilities for any mode of transportation in order to provide for the safe, efficient, rapid, comfortable, convenient, economical, and environmentally compatible movement of people and goods (transport). It is a sub-discipline of civil engineering and of industrial engineering. Transportation engineering is a major component of the civil engineering and mechanical engineering disciplines, according to specialisation of academic courses and main competences of the involved territory. The importance of transportation engineering within the civil and industrial engineering profession can be judged by the number of divisions in ASCE (American Society of Civil Engineers) that are directly related to transportation. There are six such divisions (Aerospace; Air Transportation; Highway; Pipeline; Waterway, Port, Coastal and Ocean; and Urban Transportation) representing one-third of the total 18 technical divisions within the ASCE (1987).

The planning aspects of transportation engineering relate to elements of urban planning, and involve technical forecasting decisions and political factors. Technical forecasting of passenger travel usually involves an urban transportation planning model, requiring the estimation of trip generation (how many trips for what purpose), trip distribution (destination choice, where is the traveler going), mode choice (what mode is being taken), and route assignment (which streets or routes are being used). More sophisticated forecasting can include other aspects of traveler decisions, including auto ownership, trip chaining (the decision to link individual trips together in a tour) and the choice of residential or business location (known as land use forecasting). Passenger trips are the focus of transportation engineering because they often represent the peak of demand on any transportation system.

A review of descriptions of the scope of various committees indicates that while facility planning and design continue to be the core of the transportation engineering field, such areas as operations planning, logistics, network analysis, financing, and policy analysis are also important to civil engineers, particularly to those working in highway and urban transportation. The National Council of Examiners for Engineering and Surveying (NCEES) list online the safety protocols, geometric design requirements, and signal timing.

Transportation engineering, as practiced by civil engineers, primarily involves planning, design, construction, maintenance, and operation of transportation facilities. The facilities support air, highway, railroad, pipeline, water, and even space transportation. The design aspects of transportation engineering include the sizing of transportation facilities (how many lanes or how much capacity the facility has), determining the materials and thickness used in pavement designing the geometry (vertical and horizontal alignment) of the roadway (or track).

Before any planning occurs the Engineer must take what is known as an inventory of the area or if it is appropriate, the previous system in place. This inventory or database must include information on (1)population, (2)land use, (3)economic activity, (4) transportation facilities and services, (5)travel patterns and volumes, (6)laws and ordinances, (7)regional financial resources, (8)community values and expectations. These inventories help the engineer create business models to complete accurate forecasts of the future conditions of the system Review.

Operations and management involve traffic engineering, so that vehicles move smoothly on the road or track. Older techniques include signs, signals, markings, and tolling. Newer technologies involve intelligent transportation systems, including advanced traveler information systems (such as variable message signs), advanced traffic control systems (such as ramp meters), and vehicle infrastructure integration. Human factors are an aspect of transportation engineering, particularly concerning driver-vehicle interface and user interface of road signs, signals, and markings.

Highway Engineering

Engineers in this specialization:

- Handle the planning, design, construction, and operation of highways, roads, and other vehicular facilities as well as their related bicycle and pedestrian realms.
- Estimate the transportation needs of the public and then secure the funding for the project.
- Analyze locations of high traffic volumes and high collisions for safety and capacity.
- Use civil engineering principles to improve the transportation system.
- Utilizes the three design controls which are the drivers, the vehicles, and the roadways themselves.

Railroad Engineering

Railway engineers handle the design, construction, and operation of railroads and mass transit systems that use a fixed guideway (such as light rail or even monorails). Typical tasks would include determining horizontal and vertical alignment design, station location and design, and construction cost estimating. Railroad engineers can also move into the specialized field of train dispatching which focuses on train movement control.

Railway engineers also work to build a cleaner and safer transportation network by reinvesting and revitalizing the rail system to meet future demands. In the United States, railway engineers work with elected officials in Washington, D.C. on rail transportation issues to make sure that the rail system meets the country's transportation needs.

Port and Harbor Engineering

Port and harbor engineers handle the design, construction, and operation of ports, harbors, canals, and other maritime facilities.

Airport Engineering

Airport engineers design and construct airports. Airport engineers must account for the impacts and demands of aircraft in their design of airport facilities. These engineers must use the analysis of predominant wind direction to determine runway orientation, determine the size of runway border and safety areas, different wing tip to wing tip clearances for all gates and must designate the clear zones in the entire port.

Professional Societies

- Young Professionals in Transportation (YPT)
- Institute of Transportation Engineers (ITE)
- American Society for Highway Engineering (ASHE)
- American Society of Civil Engineers (ASCE)
- The American Association of State Highway and Transportation Officials (AASHTO)
- Eastern Asia Society for Transportation Studies (EASTS)

Pavement Engineering

Pavement engineering is a branch of civil engineering that uses engineering techniques to design and maintain flexible (asphalt) and rigid (concrete) pavements. This includes streets and highways and involves knowledge of soils, hydraulics, and material properties. Pavement engineering involves new construction as well as rehabilitation and maintenance of existing pavements. Maintenance often involves using engineering judgment to make maintenance repairs with the highest long-term benefit and lowest cost. The Pavement Condition Index (PCI) is an example of an engineering approach applied to existing pavements. Another example is the use of a falling weight deflectometer (FWD) to non-destructively test existing pavements. Calculation of pavement layer strengths can be performed from the resulting deflection data. The two methods - empirical or mechanistic is used to determine pavement layer thicknesses.

References

- *Singh, Vijay P.; Ram Narayan Yadava (2003). Water Resources System Operation: Proceedings of the International Conference on Water and Environment. Allied Publishers. p. 508. ISBN 81-7764-548-X. Retrieved 2015-11-09.*
- Donald Routledge Hill (1996), "Engineering", p. 759, in *Rashed, Roshdi; Morelon, Régis (1996). Encyclopedia of the History of Arabic Science. Routledge. pp. 751–795. ISBN 0-415-12410-7*
- Adam Lucas (2006), *Wind, Water, Work: Ancient and Medieval Milling Technology*, p. 62. Brill, ISBN 90-04-14649-0.
- *Neves, edited by E. Maranha das (1991). Advances in rockfill structures. Dordrecht: Kluwer Academic. p. 341. ISBN 0-7923-1267-8. Retrieved 2015-11-09.*
- *Blight, Geoffrey E. (1998). "Construction of Tailings Dams". Case studies on tailings management. Paris, France: International Council on Metals and the Environment. pp. 9–10. ISBN 1-895720-29-X*
- Yan Ji and Stellios Plainiotis (2006): Design for Sustainability. Beijing: China Architecture and Building Press. ISBN 7-112-08390-7
- *Carter, Bill (1989). Coastal environments: an introduction to the physical, ecological, and cultural systems of coastlines. Boston: Academic Press. p. 50. ISBN 0-12-161856-0.*
- *Matzner, Richard A. (2001). Dictionary of geophysics, astrophysics, and astronomy (PDF). Boca Raton: CRC Press. p. 81. ISBN 0-8493-2891-8.*
- *Beer, Tom (1997). Environmental oceanography. Boca Raton: CRC Press. p. 44. ISBN 0-8493-8425-7. ... the reflected wave energy interacted with the incoming waves to produce standing waves known as clapotis, which promote erosion at the toe of the wall.*
- N.W.H. Allsop. (2002). Breakwaters, coastal structures and coastlines. Thomas Telford. ISBN 0-7277-3042-8.
- *"Egyptian Irrigation Bond 1898 - Aswan Dam on Nile River". Scripophily. Archived from the original on 13 May 2005. Retrieved 9 November 2015.*
- *"Bartleby.com: Great Books Online -- Quotes, Poems, Novels, Classics and hundreds more". bartleby.com. Retrieved 9 November 2015.*
- Maliva, Robert G. "Groundwater Banking: Opportunities And Management Challenges." Water Policy 16.1 (2014): 144-156. Academic Search Complete. Web. 18 Nov. 2015.
- *"Groundwater Banking - Water Education Foundation". www.watereducation.org. Retrieved 2015-09-28.*
- *O'Greene, A.T. (April 2015). "Soil suitability index identifies potential areas for groundwater banking on agricultural lands". California Agriculture.*

4

Sustainable Civil Engineering

Sustainable civil engineering is the need of the hour because of growing pollution. This chapter will discuss few of the topics which fall under the branch of sustainable civil engineering. The topics included are green building, dams, seawall, rain garden, land reclamation and permeable paving. This area is an important part of civil engineering and also plays a crucial role in environment conservation.

Dam

A dam is a barrier that impounds water or underground streams. Reservoirs created by dams not only suppress floods but also provide water for such activities as irrigation, human consumption, industrial use, aquaculture, and navigability. Hydropower is often used in conjunction with dams to generate electricity. A dam can also be used to collect water or for storage of water which can be evenly distributed between locations. Dams generally serve the primary purpose of retaining water, while other structures such as floodgates or levees (also known as dikes) are used to manage or prevent water flow into specific land regions.

A sideview of the Lake Vyrnwy dam, in Wales, finished in 1888

Karapuzha Dam, an earth dam in the Indian state of Kerala

The word *dam* can be traced back to Middle English, and before that, from Middle Dutch, as seen in the names of many old cities.

History

Ancient Dams

Early dam building took place in Mesopotamia and the Middle East. Dams were used to control the water level, for Mesopotamia's weather affected the Tigris and Euphrates rivers.

The earliest known dam is the Jawa Dam in Jordan, 100 kilometres (62 mi) northeast of the capital Amman. This gravity dam featured an originally 9-metre-high (30 ft) and 1 m-wide (3.3 ft) stone wall, supported by a 50 m-wide (160 ft) earth rampart. The structure is dated to 3000 BC.

The Ancient Egyptian Sadd-el-Kafara Dam at Wadi Al-Garawi, located about 25 km (16 mi) south of Cairo, was 102 m (335 ft) long at its base and 87 m (285 ft) wide. The structure was built around 2800 or 2600 BC as a diversion dam for flood control, but was destroyed by heavy rain during construction or shortly afterwards. During the XIIth dynasty in the 19th century BC, the Pharaohs Senosert III, Amenemhat III and Amenmehat IV dug a canal 16 km (9.9 mi) long linking the Fayum Depression to the Nile in Middle Egypt. Two dams called Ha-Uar running east-west were built to retain water during the annual flood and then release it to surrounding lands. The lake called "Mer-wer" or Lake Moeris covered 1,700 km^2 (660 sq mi) and is known today as Berkat Qaroun.

One of the engineering wonders of the ancient world was the Great Dam of Marib in Yemen. Initiated somewhere between 1750 and 1700 BC, it was made of packed earth - triangular in cross section, 580 m (1,900 ft) in length and originally 4 m (13 ft) high

- running between two groups of rocks on either side, to which it was linked by substantial stonework. Repairs were carried out during various periods, most important around 750 BC, and 250 years later the dam height was increased to 7 m (23 ft). After the end of the Kingdom of Saba, the dam fell under the control of the Ḥimyarites (~115 BC) who undertook further improvements, creating a structure 14 m (46 ft) high, with five spillway channels, two masonry-reinforced sluices, a settling pond, and a 1,000 m (3,300 ft) canal to a distribution tank. These extensive works were not actually finalized until 325 AD and allowed the irrigation of 25,000 acres (100 km^2).

By the mid-late 3rd century BC, an intricate water-management system within Dholavira in modern-day India was built. The system included 16 reservoirs, dams and various channels for collecting water and storing it.

Eflatun Pınar is a Hittite dam and spring temple near Konya, Turkey. It is thought to be from the time of the Hittite empire between the 15th and 13th century BC.

The Kallanai is constructed of unhewn stone, over 300 m (980 ft) long, 4.5 m (15 ft) high and 20 m (66 ft) wide, across the main stream of the Kaveri river in Tamil Nadu, South India. The basic structure dates to the 2nd century AD and is considered one of the oldest water-diversion or water-regulator structures in the world which is still in use. The purpose of the dam was to divert the waters of the Kaveri across the fertile delta region for irrigation via canals.

Du Jiang Yan is the oldest surviving irrigation system in China that included a dam that directed waterflow. It was finished in 251 BC. A large earthen dam, made by Sunshu Ao, the prime minister of Chu (state), flooded a valley in modern-day northern Anhui province that created an enormous irrigation reservoir (100 km (62 mi) in circumference), a reservoir that is still present today.

Roman Engineering

The Roman dam at Cornalvo in Spain has been in use for almost two millennia.

Roman dam construction was characterized by "the Romans' ability to plan and organize engineering construction on a grand scale." Roman planners introduced the then-novel concept of large reservoir dams which could secure a permanent water supply for urban settlements over the dry season. Their pioneering use of water-proof hydraulic mortar and particularly Roman concrete allowed for much larger dam structures than previously built, such as the Lake Homs Dam, possibly the largest water barrier to that date, and the Harbaqa Dam, both in Roman Syria. The highest Roman dam was the Subiaco Dam near Rome; its record height of 50 m (160 ft) remained unsurpassed until its accidental destruction in 1305.

Roman engineers made routine use of ancient standard designs like embankment dams and masonry gravity dams. Apart from that, they displayed a high degree of inventiveness, introducing most of the other basic dam designs which had been unknown until then. These include arch-gravity dams, arch dams, buttress dams and multiple arch buttress dams, all of which were known and employed by the 2nd century AD. Roman workforces also were the first to build dam bridges, such astheBridge of Valerian in Iran.

Remains of the Band-e Kaisar dam, built by the Romans in the 3rd century AD

In Iran, bridge dams such as the Band-e Kaisar were used to provide hydropower through water wheels, which often powered water-raising mechanisms. One of the first was the Roman-built dam bridge in Dezful, which could raise water 50 cubits in height for the water supply to all houses in the town. Also diversion dams were known. Milling dams were introduced which the Muslim engineers called the *Pul-i-Bulaiti*. The first was built at Shustar on the River Karun, Iran, and many of these were later built in other parts of the Islamic world. Water was conducted from the back of the dam through a large pipe to drive a water wheel and watermill. In the 10th century, Al-Muqaddasi described several dams in Persia. He reported that one in Ahwaz was more than 910 m (3,000 ft) long, and that it had many water-wheels raising the water into aqueducts through which it flowed into reservoirs of the city. Another one, the Band-i-Amir dam, provided irrigation for 300 villages.

Middle Ages

In the Netherlands, a low-lying country, dams were often applied to block rivers in order to regulate the water level and to prevent the sea from entering the marsh lands. Such dams often marked the beginning of a town or city because it was easy to cross the river at such a place, and often gave rise to the respective place's names in Dutch.

For instance the Dutch capital Amsterdam (old name *Amstelredam*) started with a dam through the river Amstel in the late 12th century, and Rotterdam started with a dam through the river Rotte, a minor tributary of the Nieuwe Maas. The central square of Amsterdam, covering the original place of the 800-year-old dam, still carries the name *Dam Square* or simply *the Dam*.

Industrial Revolution

An engraving of the Rideau Canal locks at Bytown

The Romans were the first to build arch dams, where the reaction forces from the abutment stabilizes the structure from the external hydrostatic pressure, but it was only in the 19th century that the engineering skills and construction materials available were capable of building the first large-scale arch dams.

Three pioneering arch dams were built around the British Empire in the early 19th century. Henry Russel of the Royal Engineers oversaw the construction of the Mir Alam dam in 1804 to supply water to the city of Hyderabad (it is still in use today). It had a height of 12 m (39 ft) and consisted of 21 arches of variable span.

In the 1820s and 30s, Lieutenant-Colonel John By supervised the construction of the Rideau Canal in Canada near modern-day Ottawa and built a series of curved masonry dams as part of the waterway system. In particular, the Jones Falls Dam, built by John Redpath, was completed in 1832 as the largest dam in North America and an engineering marvel. In order to keep the water in control during construction, two sluices, artificial channels for conducting water, were kept open in the dam. The first was near the base of the dam on its east side. A second sluice was put in on the west side of the dam, about 20 ft (6.1 m) above the base. To make the switch from the lower to upper sluice, the outlet of Sand Lake was blocked off.

Masonry arch wall, Parramatta, New South Wales, the first engineered dam built in Australia

Hunts Creek near the city of Parramatta, Australia, was dammed in the 1850s, to cater for the demand for water from the growing population of the city. The masonry arch dam wall was designed by Lieutenant Percy Simpson who was influenced by the advances in dam engineering techniques made by the Royal Engineers in India. The dam cost £17,000 and was completed in 1856 as the first engineered dam built in Australia, and the second arch dam in the world built to mathematical specifications.

The first such dam was opened two years earlier in France. It was the first French arch dam of the industrial era, and it was built by François Zola in the municipality of Aix-en-Provence to improve the supply of water after the 1832 cholera outbreak devastated the area. After royal approval was granted in 1844, the dam was constructed over the following decade. Its construction was carried out on the basis of the mathematical results of scientific stress analysis.

The 75-miles dam near Warwick, Australia, was possibly the world's first concrete arch dam. Designed by Henry Charles Stanley in 1880 with an overflow spillway and a special water outlet, it was eventually heightened to 10 m (33 ft).

In the latter half of the nineteenth century, significant advances in the scientific theory of masonry dam design were made. This transformed dam design from an art based on empirical methodology to a profession based on a rigorously applied scientific theoretical framework. This new emphasis was centered around the engineering faculties of universities in France and in the United Kingdom. William John Macquorn Rankine at the University of Glasgow pioneered the theoretical understanding of dam structures in his 1857 paper *On the Stability of Loose Earth*. Rankine theory provided a good understanding of the principles behind dam design. In France, J. Augustin Tortene de Sazilly explained the mechanics of vertically faced masonry gravity dams, and Zola's dam was the first to be built on the basis of these principles.

Large Dams

The Hoover Dam by Ansel Adams, 1942

The era of large dams was initiated with the construction of the Aswan Low Dam in Egypt in 1902, a gravity masonry buttress dam on the Nile River. Following their 1882 invasion and occupation of Egypt, the British began construction in 1898. The project was designed by Sir William Willcocks and involved several eminent engineers of the time, including Sir Benjamin Baker and Sir John Aird, whose firm, John Aird & Co., was the main contractor. Capital and financing were furnished by Ernest Cassel. When initially constructed between 1899 and 1902, nothing of its scale had ever been attempted; on completion, it was the largest masonry dam in the world.

The Hoover Dam is a massive concrete arch-gravity dam, constructed in the Black Canyon of the Colorado River, on the border between the US states of Arizona and Nevada between 1931 and 1936 during the Great Depression. In 1928, Congress authorized the project to build a dam that would control floods, provide irrigation water and produce hydroelectric power. The winning bid to build the dam was submitted by a consortium called Six Companies, Inc. Such a large concrete structure had never been built before, and some of the techniques were unproven. The torrid summer weather and the lack of facilities near the site also presented difficulties. Nevertheless, Six Companies turned over the dam to the federal government on 1 March 1936, more than two years ahead of schedule.

By 1997, there were an estimated 800,000 dams worldwide, some 40,000 of them over 15 m (49 ft) high. In 2014, scholars from the University of Oxford published a study of the cost of large dams – based on the largest existing dataset – documenting significant cost overruns for a majority of dams and questioning whether benefits typically offset costs for such dams.

Types of Dams

Dams can be formed by human agency, natural causes, or even by the intervention of wildlife such as beavers. Man-made dams are typically classified according to their size (height), intended purpose or structure.

By Structure

Based on structure and material used, dams are classified as easily created without materials, arch-gravity dams, embankment dams or masonry dams, with several subtypes.

Arch Dams

Gordon Dam, Tasmania, is an arch dam.

In the arch dam, stability is obtained by a combination of arch and gravity action. If the upstream face is vertical the entire weight of the dam must be carried to the foundation by gravity, while the distribution of the normal hydrostatic pressure between vertical cantilever and arch action will depend upon the stiffness of the dam in a vertical and horizontal direction. When the upstream face is sloped the distribution is more complicated. The normal component of the weight of the arch ring may be taken by the arch action, while the normal hydrostatic pressure will be distributed as described above.

For this type of dam, firm reliable supports at the abutments (either buttress or canyon side wall) are more important. The most desirable place for an arch dam is a narrow canyon with steep side walls composed of sound rock. The safety of an arch dam is dependent on the strength of the side wall abutments, hence not only should the arch be well seated on the side walls but also the character of the rock should be carefully inspected.

Daniel-Johnson Dam, Quebec, is a multiple-arch buttress dam.

Two types of single-arch dams are in use, namely the constant-angle and the constant-radius dam. The constant-radius type employs the same face radius at all elevations of the dam, which means that as the channel grows narrower towards the bottom of the dam the central angle subtended by the face of the dam becomes smaller. Jones Falls Dam, in Canada, is a constant radius dam. In a constant-angle dam, also known as a variable radius dam, this subtended angle is kept a constant and the variation in distance between the abutments at various levels are taken care of by varying the radii. Constant-radius dams are much less common than constant-angle dams. Parker Dam on the Colorado River is a constant-angle arch dam.

A similar type is the double-curvature or thin-shell dam. Wildhorse Dam near Mountain City, Nevada, in the United States is an example of the type. This method of construction minimizes the amount of concrete necessary for construction but transmits large loads to the foundation and abutments. The appearance is similar to a single-arch dam but with a distinct vertical curvature to it as well lending it the vague appearance of a concave lens as viewed from downstream.

The multiple-arch dam consists of a number of single-arch dams with concrete buttresses as the supporting abutments, as for example the Daniel-Johnson Dam, Québec, Canada. The multiple-arch dam does not require as many buttresses as the hollow gravity type, but requires good rock foundation because the buttress loads are heavy.

Gravity Dams

The Grand Coulee Dam is an example of a solid gravity dam.

In a gravity dam, the force that holds the dam in place against the push from the water is Earth's gravity pulling down on the mass of the dam. The water presses laterally (downstream) on the dam, tending to overturn the dam by rotating about its toe (a point at the bottom downstream side of the dam). The dam's weight counteracts that force, tending to rotate the dam the other way about its toe. The designer ensures that the dam is heavy enough that the dam's weight wins that contest. In engineering terms, that is true whenever the resultant of the forces of gravity acting on the dam and water pressure on the dam acts in a line that passes upstream of the toe of the dam.

Furthermore, the designer tries to shape the dam so if one were to consider the part of dam above any particular height to be a whole dam itself, that dam also would be held in place by gravity. i.e. there is no tension in the upstream face of the dam holding the top of the dam down. The designer does this because it is usually more practical to make a dam of material essentially just piled up than to make the material stick together against vertical tension.

Note that the shape that prevents tension in the upstream face also eliminates a balancing compression stress in the downstream face, providing additional economy.

For this type of dam, it is essential to have an impervious foundation with high bearing strength.

When situated on a suitable site, a gravity dam can prove to be a better alternative to other types of dams. When built on a carefully studied foundation, the gravity dam probably represents the best developed example of dam building. Since the fear of flood is a strong motivator in many regions, gravity dams are being built in some instances where an arch dam would have been more economical.

Gravity dams are classified as "solid" or "hollow" and are generally made of either concrete or masonry. The solid form is the more widely used of the two, though the hollow dam is frequently more economical to construct. Grand Coulee Dam is a solid gravity dam and Braddock Locks & Dam is a hollow gravity dam.

Arch-Gravity Dams

The Hoover Dam is an example of an arch-gravity dam.

A gravity dam can be combined with an arch dam into an arch-gravity dam for areas with massive amounts of water flow but less material available for a purely gravity dam. The inward compression of the dam by the water reduces the lateral (horizontal) force acting on the dam. Thus, the gravitation force required by the dam is lessened, i.e. the dam does not need to be so massive. This enables thinner dams and saves resources.

Barrages

The Koshi Barrage

A barrage dam is a special kind of dam which consists of a line of large gates that can be opened or closed to control the amount of water passing the dam. The gates are set between flanking piers which are responsible for supporting the water load, and are often used to control and stabilize water flow for irrigation systems. An example of this type of dam is the now-decommissioned Red Bluff Diversion Dam on the Sacramento River near Red Bluff, California.

Barrages that are built at the mouths of rivers or lagoons to prevent tidal incursions or utilize the tidal flow for tidal power are known as tidal barrages.

Embankment Dams

Embankment dams are made from compacted earth, and have two main types, rock-fill and earth-fill dams. Embankment dams rely on their weight to hold back the force of water, like gravity dams made from concrete.

Rock-Fill Dams

The Gathright Dam in Virginia is a rock-fill embankment dam.

Rock-fill dams are embankments of compacted free-draining granular earth with an impervious zone. The earth utilized often contains a high percentage of large particles, hence the term "rock-fill". The impervious zone may be on the upstream face and made of masonry, concrete, plastic membrane, steel sheet piles, timber or other material. The impervious zone may also be within the embankment in which case it is referred to as a *core*. In the instances where clay is utilized as the impervious material the dam is referred to as a *composite* dam. To prevent internal erosion of clay into the rock fill due to seepage forces, the core is separated using a filter. Filters are specifically graded soil designed to prevent the migration of fine grain soil particles. When suitable material is at hand, transportation is minimized leading to cost savings during construction. Rock-fill dams are resistant to damage from earthquakes. However, inadequate quality control during construction can lead to poor compaction and sand in the embankment which can lead to liquefaction of the rock-fill during an earthquake. Liquefaction potential can be reduced by keeping susceptible material from being saturated, and by providing adequate compaction during construction. An example of a rock-fill dam is New Melones Dam in California or the Fierza Dam in Albania.

A core that is growing in popularity is asphalt concrete. The majority of such dams are built with rock and/or gravel as the main fill material. Almost 100 dams of this design have now been built worldwide since the first such dam was completed in 1962. All asphalt-concrete core dams built so far have an excellent performance record. The type of asphalt used is a viscoelastic-plastic material that can adjust to the movements and deformations imposed on the embankment as a whole, and to settlements in the foundation. The flexible properties of the asphalt make such dams especially suited in earthquake regions.

For the Moglicë Hydro Power Plant in Albania the Norwegian power company Statkraft is currently building an asphalt-core rock-fill dam. Upon completion in 2018 the 320 m long, 150 m high and 460 m wide dam is anticipated to be the world's highest of its kind.

Concrete-Face Rock-Fill Dams

A concrete-face rock-fill dam (CFRD) is a rock-fill dam with concrete slabs on its upstream face. This design provides the concrete slab as an impervious wall to prevent leakage and also a structure without concern for uplift pressure. In addition, the CFRD design is flexible for topography, faster to construct and less costly than earth-fill dams. The CFRD concept originated during the California Gold Rush in the 1860s when miners constructed rock-fill timber-face dams for sluice operations. The timber was later replaced by concrete as the design was applied to irrigation and power schemes. As CFRD designs grew in height during the 1960s, the fill was compacted and the slab's horizontal and vertical joints were replaced with improved vertical joints. In the last few decades, the design has become popular.

Currently, the tallest CFRD in the world is the 233 m-tall (764 ft) Shuibuya Dam in China which was completed in 2008.

Earth-Fill Dams

Earth-fill dams, also called earthen dams, rolled-earth dams or simply earth dams, are constructed as a simple embankment of well compacted earth. A homogeneous rolled-earth dam is entirely constructed of one type of material but may contain a drain layer to collect seep water. A zoned-earth dam has distinct parts or zones of dissimilar material, typically a locally plentiful shell with a watertight clay core. Modern zoned-earth embankments employ filter and drain zones to collect and remove seep water and preserve the integrity of the downstream shell zone. An outdated method of zoned earth dam construction utilized a hydraulic fill to produce a watertight core. Rolled-earth dams may also employ a watertight facing or core in the manner of a rock-fill dam. An interesting type of temporary earth dam occasionally used in high latitudes is the frozen-core dam, in which a coolant is circulated through pipes inside the dam to maintain a watertight region of permafrost within it.

Tarbela Dam is a large dam on the Indus River in Pakistan. It is located about 50 km (31 mi) northwest of Islamabad, and a height of 485 ft (148 m) above the river bed and a reservoir size of 95 sq mi (250 km^2) makes it the largest earth-filled dam in the world. The principal element of the project is an embankment 9,000 feet (2,700 m) long with a maximum height of 465 feet (142 m). The total volume of earth and rock used for the project is approximately 200 million cubic yards (152.8 million cu. meters) which makes it one of the largest man-made structures in the world.

Because earthen dams can be constructed from materials found on-site or nearby, they can be very cost-effective in regions where the cost of producing or bringing in concrete would be prohibitive.

By Size

International standards (including the International Commission on Large Dams, ICOLD) define *large dams* as higher than 15 m (49 ft) and *major dams* as over 150 m (490 ft) in height. The *Report of the World Commission on Dams* also includes in the *large* category, dams, such as barrages, which are between 5 and 15 m (16 and 49 ft) high with a reservoir capacity of more than 3 million cubic metres (2,400 acre·ft).

The tallest dam in the world is the 300 m-high (980 ft) Nurek Dam in Tajikistan.

By Use

Saddle Dam

A saddle dam is an auxiliary dam constructed to confine the reservoir created by a primary dam either to permit a higher water elevation and storage or to limit the extent of a reservoir for increased efficiency. An auxiliary dam is constructed in a low spot or "saddle" through which the reservoir would otherwise escape. On occasion, a reservoir is contained by a similar structure called a dike to prevent inundation of nearby land. Dikes are commonly used for reclamation of arable land from a shallow lake. This is similar to a levee, which is a wall or embankment built along a river or stream to protect adjacent land from flooding.

Weir

A weir (also sometimes called an *overflow dam*) is a type of small overflow dam that is often used within a river channel to create an impoundment lake for water abstraction purposes and which can also be used for flow measurement or retardation.

Check Dam

A check dam is a small dam designed to reduce flow velocity and control soil erosion. Conversely, a wing dam is a structure that only partly restricts a waterway, creating a faster channel that resists the accumulation of sediment.

Dry Dam

A dry dam, also known as a flood retarding structure, is a dam designed to control flooding. It normally holds back no water and allows the channel to flow freely, except during periods of intense flow that would otherwise cause flooding downstream.

Diversionary Dam

A diversionary dam is a structure designed to divert all or a portion of the flow of a river from its natural course. The water may be redirected into a canal or tunnel for irrigation and/or hydroelectric power production.

Underground Dam

Underground dams are used to trap groundwater and store all or most of it below the surface for extended use in a localized area. In some cases they are also built to prevent saltwater from intruding into a freshwater aquifer. Underground dams are typically constructed in areas where water resources are minimal and need to be efficiently stored, such as in deserts and on islands like the Fukuzato Dam in Okinawa, Japan. They are most common in northeastern Africa and the arid areas of Brazil while also being used in the southwestern United States, Mexico, India, Germany, Italy, Greece, France and Japan.

There are two types of underground dams: a *sub-surface* and a *sand-storage* dam. A sub-surface dam is built across an aquifer or drainage route from an impervious layer (such as solid bedrock) up to just below the surface. They can be constructed of a variety of materials to include bricks, stones, concrete, steel or PVC. Once built, the water stored behind the dam raises the water table and is then extracted with wells. A sand-storage dam is a weir built in stages across a stream or wadi. It must be strong, as floods will wash over its crest. Over time, sand accumulates in layers behind the dam, which helps store water and, most importantly, prevent evaporation. The stored water can be extracted with a well, through the dam body, or by means of a drain pipe.

Tailings Dam

A tailings dam is typically an earth-fill embankment dam used to store tailings, which are produced during mining operations after separating the valuable fraction from the uneconomic fraction of an ore. Conventional water retention dams can serve this purpose, but due to cost, a tailings dam is more viable. Unlike water retention dams, a tailings dam is raised in succession throughout the life of the particular mine. Typically, a base or starter dam is constructed, and as it fills with a mixture of tailings and water, it is raised. Material used to raise the dam can include the tailings (depending on their size) along with dirt.

There are three raised tailings dam designs, the *upstream*, *downstream* and *centerline*, named according to the movement of the crest during raising. The specific design used is dependent upon topography, geology, climate, the type of tailings, and cost. An upstream tailings dam consists of trapezoidal embankments being constructed on top but toe to crest of another, moving the crest further upstream. This creates a relatively flat downstream side and a jagged upstream side which is supported by tailings slurry in the impoundment. The downstream design refers to the successive raising of the embankment that positions the fill and crest further downstream. A centerlined dam has sequential embankment dams constructed directly on top of another while fill is placed on the downstream side for support and slurry supports the upstream side.

Because tailings dams often store toxic chemicals from the mining process, they have an impervious liner to prevent seepage. Water/slurry levels in the tailings pond must be managed for stability and environmental purposes as well.

By Material

Steel Dams

Red Ridge steel dam, built 1905, Michigan

A steel dam is a type of dam briefly experimented with around the start of the 20th century which uses steel plating (at an angle) and load-bearing beams as the structure. Intended as permanent structures, steel dams were an (arguably failed) experiment to determine if a construction technique could be devised that was cheaper than masonry, concrete or earthworks, but sturdier than timber crib dams.

Timber Dams

A timber crib dam in Michigan, photographed in 1978

Timber dams were widely used in the early part of the industrial revolution and in frontier areas due to ease and speed of construction. Rarely built in modern times because of their relatively short lifespan and the limited height to which they can be built, timber dams must be kept constantly wet in order to maintain their water retention properties and limit deterioration by rot, similar to a barrel. The locations where timber dams are most economical to build are those where timber is plentiful, cement is costly or difficult to transport, and either a low head diversion dam is required or longevity is not an issue. Timber dams were once numerous, especially in the North American West, but most have failed, been hidden under earth embankments, or been replaced with entirely new structures. Two common variations of timber dams were the *crib* and the *plank*.

Timber crib dams were erected of heavy timbers or dressed logs in the manner of a log house and the interior filled with earth or rubble. The heavy crib structure supported the dam's face and the weight of the water. Splash dams were timber crib dams used to help float logs downstream in the late 19th and early 20th centuries.

Timber plank dams were more elegant structures that employed a variety of construction methods utilizing heavy timbers to support a water retaining arrangement of planks.

Other Types

Cofferdams

A cofferdam during the construction of locks at the Montgomery Point Lock and Dam

A cofferdam is a barrier, usually temporary, constructed to exclude water from an area that is normally submerged. Made commonly of wood, concrete, or steel sheet piling, cofferdams are used to allow construction on the foundation of permanent dams, bridges, and similar structures. When the project is completed, the cofferdam will usually be demolished or removed unless the area requires continuous maintenance.

Common uses for cofferdams include construction and repair of offshore oil platforms. In such cases the cofferdam is fabricated from sheet steel and welded into place under water. Air is pumped into the space, displacing the water and allowing a dry work environment below the surface.

Natural Dams

Dams can also be created by natural geological forces. Volcanic dams are formed when lava flows, often basaltic, intercept the path of a stream or lake outlet, resulting in the creation of a natural impoundment. An example would be the eruptions of the Uinkaret volcanic field about 1.8 million–10,000 years ago, which created lava dams on the Colorado River in northern Arizona in the United States. The largest such lake grew to about 800 km (500 mi) in length before the failure of its dam. Glacial activity can also form natural dams, such as the damming of the Clark Fork in Montana by the Cordilleran Ice Sheet, which formed the 7,780 km^2 (3,000 sq mi) Glacial Lake Missoula near the end of the last Ice Age. Moraine deposits left behind by glaciers can also dam rivers to form lakes, such as at Flathead Lake, also in Montana.

Natural disasters such as earthquakes and landslides frequently create landslide dams in mountainous regions with unstable local geology. Historical examples include the Usoi Dam in Tajikistan, which blocks the Murghab River to create Sarez Lake. At 560 m (1,840 ft) high, it is the tallest dam in the world, including both natural and man-made dams. A more recent example would be the creation of Attabad Lake by a landslide on Pakistan's Hunza River.

Natural dams often pose significant hazards to human settlements and infrastructure. The resulting lakes often flood inhabited areas, while a catastrophic failure of the dam could cause even greater damage, such as the failure of western Wyoming's Gros Ventre landslide dam in 1927, which wiped out the town of Kelly and resulted in the deaths of six people.

Beaver Dams

Beavers create dams primarily out of mud and sticks to flood a particular habitable area. By flooding a parcel of land, beavers can navigate below or near the surface and remain relatively well hidden or protected from predators. The flooded region also allows beavers access to food, especially during the winter.

Construction Elements

Power Generation Plant

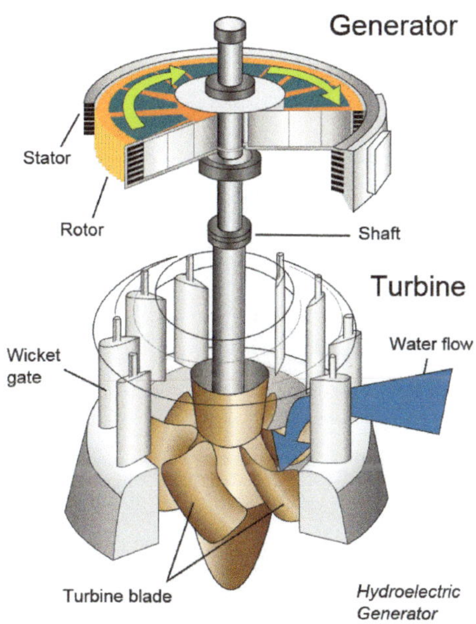

Hydraulic turbine and electric generator

As of 2005, hydroelectric power, mostly from dams, supplies some 19% of the world's electricity, and over 63% of renewable energy. Much of this is generated by large dams, although China uses small-scale hydro generation on a wide scale and is responsible for about 50% of world use of this type of power.

Most hydroelectric power comes from the potential energy of dammed water driving a water turbine and generator; to boost the power generation capabilities of a dam, the water may be run through a large pipe called a penstock before the turbine. A variant on this simple model uses pumped-storage hydroelectricity to produce electricity to match periods of high and low demand, by moving water between reservoirs at different elevations. At times of low electrical demand, excess generation capacity is used to pump water into the higher reservoir. When there is higher demand, water is released back into the lower reservoir through a turbine.

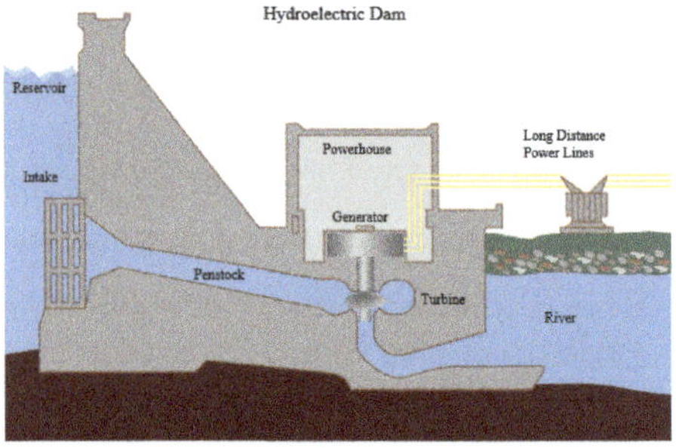

Hydroelectric dam in cross section

Spillways

Spillway on Llyn Brianne dam, Wales, soon after first fill

A spillway is a section of a dam designed to pass water from the upstream side of a dam to the downstream side. Many spillways have floodgates designed to control the flow through the spillway. There are several types of spillway. A *service spillway* or *primary spillway* passes normal flow. An *auxiliary spillway* releases flow in excess of the capacity of the service spillway. An *emergency spillway* is designed for extreme conditions, such as a serious malfunction of the service spillway. A *fuse plug spillway* is a low embankment designed to be overtopped and washed away in the event of a large flood. The elements of a fuse plug are independent free-standing blocks, set side by side which work without any remote control. They allow increasing the normal pool of the dam without compromising the security of the dam because they are designed to be gradually evacuated for exceptional events. They work as fixed weirs at times by allowing over-flow for common floods.

The spillway can be gradually eroded by water flow, including cavitation or turbulence of the water flowing over the spillway, leading to its failure. It was the inadequate design of the spillway which led to the 1889 over-topping of the South Fork Dam in Johnstown, Pennsylvania, resulting in the infamous Johnstown Flood (the "great flood of 1889").

Erosion rates are often monitored, and the risk is ordinarily minimized, by shaping the downstream face of the spillway into a curve that minimizes turbulent flow, such as an ogee curve.

Dam Creation

Common Purposes

Function	Example
Power generation	Hydroelectric power is a major source of electricity in the world. Many countries have rivers with adequate water flow, that can be dammed for power generation purposes. For example, the Itaipu Dam on the Paraná River in South America generates 14 GW and supplied 93% of the energy consumed by Paraguay and 20% of that consumed by Brazil as of 2005.
Water supply	Many urban areas of the world are supplied with water abstracted from rivers pent up behind low dams or weirs. Examples include London, with water from the River Thames, and Chester, with water taken from the River Dee. Other major sources include deep upland reservoirs contained by high dams across deep valleys, such as the Claerwen series of dams and reservoirs.
Stabilize water flow / irrigation	Dams are often used to control and stabilize water flow, often for agricultural purposes and irrigation. Others such as the Berg Strait dam can help to stabilize or restore the water levels of inland lakes and seas, in this case the Aral Sea.
Flood prevention	The Keenleyside Dam on the Columbia River, Canada can store 8.76 km^3 (2.10 cu mi) of floodwaters, and the huge Delta Works protects the Netherlands from coastal flooding.

Land reclamation	Dams (often called dykes or levees in this context) are used to prevent ingress of water to an area that would otherwise be submerged, allowing its reclamation for human use.
Water diversion	A typically small dam used to divert water for irrigation, power generation, or other uses, with usually no other function. Occasionally, they are used to divert water to another drainage or reservoir to increase flow there and improve water use in that particular area.
Navigation	Dams create deep reservoirs and can also vary the flow of water downstream. This can in return affect upstream and downstream navigation by altering the river's depth. Deeper water increases or creates freedom of movement for water vessels. Large dams can serve this purpose, but most often weirs and locks are used.

Some of these purposes are conflicting, and the dam operator needs to make dynamic tradeoffs. For example, power generation and water supply would keep the reservoir high, whereas flood prevention would keep it low. Many dams in areas where precipitation fluctuates in an annual cycle will also see the reservoir fluctuate annually in an attempt to balance these difference purposes. Dam management becomes a complex exercise amongst competing stakeholders.

Location

The discharge of Takato Dam

One of the best places for building a dam is a narrow part of a deep river valley; the valley sides then can act as natural walls. The primary function of the dam's structure is to fill the gap in the natural reservoir line left by the stream channel. The sites are usually those where the gap becomes a minimum for the required storage capacity. The most economical arrangement is often a composite structure such as a masonry dam flanked by earth embankments. The current use of the land to be flooded should be dispensable.

Significant other engineering and engineering geology considerations when building a dam include:

- Permeability of the surrounding rock or soil
- Earthquake faults
- Landslides and slope stability
- Water table
- Peak flood flows
- Reservoir silting
- Environmental impacts on river fisheries, forests and wildlife
- Impacts on human habitations
- Compensation for land being flooded as well as population resettlement
- Removal of toxic materials and buildings from the proposed reservoir area

Impact Assessment

Impact is assessed in several ways: the benefits to human society arising from the dam (agriculture, water, damage prevention and power), harm or benefit to nature and wildlife, impact on the geology of an area (whether the change to water flow and levels will increase or decrease stability), and the disruption to human lives (relocation, loss of archeological or cultural matters underwater).

Environmental Impact

Wood and garbage accumulation due to a dam

Reservoirs held behind dams affect many ecological aspects of a river. Rivers topography and dynamics depend on a wide range of flows, whilst rivers below dams often experience long periods of very stable flow conditions or sawtooth flow patterns caused by releases followed by no releases. Water releases from a reservoir including that exiting a turbine usually contain very little suspended sediment, and this in turn can lead to scouring of river beds and loss of riverbanks; for example, the daily cyclic flow variation caused by the Glen Canyon Dam was a contributor to sand bar erosion.

Older dams often lack a fish ladder, which keeps many fish from moving upstream to their natural breeding grounds, causing failure of breeding cycles or blocking of migration paths. Even the presence of a fish ladder does not always prevent a reduction in fish reaching the spawning grounds upstream. In some areas, young fish ("smolt") are transported downstream by barge during parts of the year. Turbine and power-plant designs that have a lower impact upon aquatic life are an active area of research.

A large dam can cause the loss of entire ecospheres, including endangered and undiscovered species in the area, and the replacement of the original environment by a new inland lake.

Large reservoirs formed behind dams have been indicated in the contribution of seismic activity, due to changes in water load and/or the height of the water table.

Dams are also found to have a role in the increase/decrease of global warming. The changing water levels in reservoirs are a source for greenhouse gases like methane. While dams and the water behind them cover only a small portion of earth's surface, they harbour biological activity that can produce large amounts of greenhouse gases.

Human Social Impact

The impact on human society is also significant. Nick Cullather argues in *Hungry World: America's Cold War Battle Against Poverty in Asia* that dam construction requires the state to displace individual people in the name of the common good, and that it often leads to abuses of the masses by planners. He cites Morarji Desai, Interior Minister of India, in 1960 speaking to villagers upset about the Pong Dam, who threatened to "release the waters" and drown the villagers if they did not cooperate.

For example, the Three Gorges Dam on the Yangtze River in China is more than five times the size of the Hoover Dam (U.S.), and creates a reservoir 600 km (370 mi) long to be used for flood control and hydro-power generation. Its construction required the loss of over a million people's homes and their mass relocation, the loss of many valuable archaeological and cultural sites, as well as significant ecological change. During the 2010 China floods, the dam held back a what would have been a disastrous flood and the huge reservoir rose by 4 m (13 ft) overnight.

It is estimated that to date, 40–80 million people worldwide have been physically displaced from their homes as a result of dam construction.

Economics

Construction of a hydroelectric plant requires a long lead time for site studies, hydrological studies, and environmental impact assessments, and are large-scale projects by comparison to traditional power generation based upon fossil fuels. The number of sites that can be economically developed for hydroelectric production is limited; new sites tend to be far from population centers and usually require extensive power transmission lines. Hydroelectric generation can be vulnerable to major changes in the climate, including variations in rainfall, ground and surface water levels, and glacial melt, causing additional expenditure for the extra capacity to ensure sufficient power is available in low-water years.

Once completed, if it is well designed and maintained, a hydroelectric power source is usually comparatively cheap and reliable. It has no fuel and low escape risk, and as an alternative energy source it is cheaper than both nuclear and wind power. It is more easily regulated to store water as needed and generate high power levels on demand compared to wind power.

Dam Failure

The reservoir emptying through the failed Teton Dam

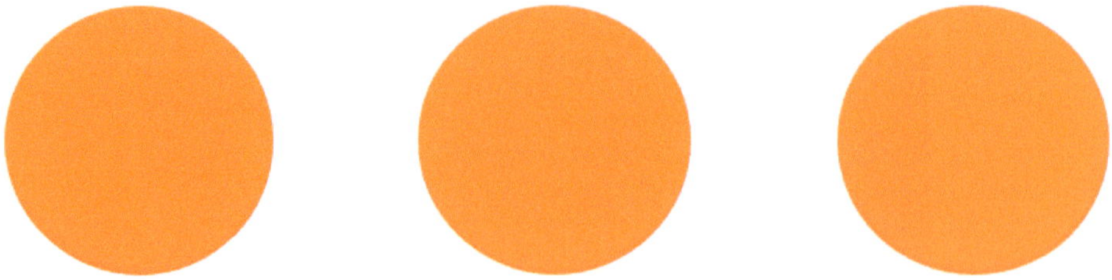

International special sign for works and installations containing dangerous forces

Dam failures are generally catastrophic if the structure is breached or significantly damaged. Routine deformation monitoring and monitoring of seepage from drains in and around larger dams is useful to anticipate any problems and permit remedial action to be taken before structural failure occurs. Most dams incorporate mechanisms to permit the reservoir to be lowered or even drained in the event of such problems. Another solution can be rock grouting – pressure pumping portland cement slurry into weak fractured rock.

During an armed conflict, a dam is to be considered as an "installation containing dangerous forces" due to the massive impact of a possible destruction on the civilian population and the environment. As such, it is protected by the rules of international humanitarian law (IHL) and shall not be made the object of attack if that may cause severe losses among the civilian population. To facilitate the identification, a protective sign consisting of three bright orange circles placed on the same axis is defined by the rules of IHL.

The main causes of dam failure include inadequate spillway capacity, piping through the embankment, foundation or abutments, spillway design error (South Fork Dam), geological instability caused by changes to water levels during filling or poor surveying (Vajont, Malpasset, Testalinden Creek dams), poor maintenance, especially of outlet pipes (Lawn Lake Dam, Val di Stava Dam collapse), extreme rainfall (Shakidor Dam), earthquakes, and human, computer or design error (Buffalo Creek Flood, Dale Dike Reservoir, Taum Sauk pumped storage plant).

A notable case of deliberate dam failure (prior to the above ruling) was the Royal Air Force 'Dambusters' raid on Germany in World War II (codenamed "Operation Chastise"), in which three German dams were selected to be breached in order to damage German infrastructure and manufacturing and power capabilities deriving from the Ruhr and Eder rivers. This raid later became the basis for several films.

Since 2007, the Dutch IJkdijk foundation is developing, with an open innovation model and early warning system for levee/dike failures. As a part of the development effort, full-scale dikes are destroyed in the IJkdijk fieldlab. The destruction process is monitored by sensor networks from an international group of companies and scientific institutions.

Green Building

US EPA Kansas City Science & Technology Center This facility features the following green attributes:

- LEED 2.0 Gold certified
- Green Power
- Native Landscaping

Green building (also known as green construction or sustainable building) refers to both a structure and the using of processes that are environmentally responsible and resource-efficient throughout a building's life-cycle: from siting to design, construction, operation, maintenance, renovation, and demolition. In other words, green building design involves finding the balance between homebuilding and the sustainable environment. This requires close cooperation of the design team, the architects, the engineers, and the client at all project stages. The Green Building practice expands and complements the classical building design concerns of economy, utility, durability, and comfort.

Leadership in Energy and Environmental Design (LEED) is a set of rating systems for the design, construction, operation, and maintenance of green buildings which was Developed by the U.S. Green Building Council. Other certificates system that confirms the sustainability of buildings is the British BREEAM (Building Research Establishment Environmental Assessment Method) for buildings and large scale developments. Currently, World Green Building Council is conducting research on the effects of green buildings on the health and productivity of their users and is working with World Bank to promote Green Buildings in Emerging Markets through EDGE Excellence in Design for Greater Efficiencies Market Transformation Program and certification.

Although new technologies are constantly being developed to complement current practices in creating greener structures, the common objective of green buildings is to reduce the overall impact of the built environment on human health and the natural environment by:

- Efficiently using energy, water, and other resources
- Protecting occupant health and improving employee productivity
- Reducing waste, pollution and environmental degradation

A similar concept is natural building, which is usually on a smaller scale and tends to focus on the use of natural materials that are available locally. Other related topics include sustainable design and green architecture. Sustainability may be defined as meeting the needs of present generations without compromising the ability of future generations to meet their needs. Although some green building programs don't address the issue of the retrofitting existing homes, others do, especially through public schemes for energy efficient refurbishment. Green construction principles can easily be applied to retrofit work as well as new construction.

A 2009 report by the U.S. General Services Administration found 12 sustainably-designed buildings that cost less to operate and have excellent energy performance. In addition, occupants were overall more satisfied with the building than those in typical commercial buildings.These are eco-friendly buildings.

Reducing Environmental Impact

Hanging gardens of One Central Park, Sydney

Globally buildings are responsible for a huge share of energy, electricity, water and materials consumption. The building sector has the greatest potential to deliver significant cuts in emissions at little or no cost. Buildings account for 18% of global emissions today, or the equivalent of 9 billion tonnes of CO_2 annually. If new technologies in construction are not adopted during this time of rapid growth, emissions could double by 2050, according to the United Nations Environment Program.Green building practices aim to reduce the environmental impact of building. Since construction almost always degrades a building site, not building at all is preferable to green building, in terms of reducing environmental impact. The second rule is that every building should be as small as possible. The third rule is not to contribute to sprawl, even if the most energy-efficient, environmentally sound methods are used in design and construction.

Buildings account for a large amount of land. According to the National Resources Inventory, approximately 107 million acres (430,000 km²) of land in the United States are developed. The International Energy Agency released a publication that estimated that existing buildings are responsible for more than 40% of the world's total primary energy consumption and for 24% of global carbon dioxide emissions.

Goals of Green Building

Blu Homes mkSolaire, a green building designed by Michelle Kaufmann.

The concept of sustainable development can be traced to the energy (especially fossil oil) crisis and environmental pollution concerns of the 1960s and 1970s. The Rachel Carson book, "Silent Spring", published in 1962, is considered to be one of the first initial efforts to describe sustainable development as related to green building. The green building movement in the U.S. originated from the need and desire for more energy efficient and environmentally friendly construction practices. There are a number of motives for building green, including environmental, economic, and social benefits. However, modern sustainability initiatives call for an integrated and synergistic design to both new construction and in the retrofitting of existing structures. Also known as sustainable design, this approach integrates the building life-cycle with each green practice employed with a design-purpose to create a synergy among the practices used.

Taipei 101, the tallest and largest green building of LEED Platinum certification in the world since 2011.

Green building brings together a vast array of practices, techniques, and skills to reduce and ultimately eliminate the impacts of buildings on the environment and human health. It often emphasizes taking advantage of renewable resources, e.g., using sunlight through passive solar, active solar, and photovoltaic equipment, and using plants and trees through green roofs, rain gardens, and reduction of rainwater run-off. Many other techniques are used, such as using low-impact building materials or using packed gravel or permeable concrete instead of conventional concrete or asphalt to enhance replenishment of ground water.

While the practices or technologies employed in green building are constantly evolving and may differ from region to region, fundamental principles persist from which the method is derived: siting and structure design efficiency, energy efficiency, water efficiency, materials efficiency, indoor environmental quality enhancement, operations and maintenance optimization and waste and toxics reduction. The essence of green building is an optimization of one or more of these principles. Also, with the proper synergistic design, individual green building technologies may work together to produce a greater cumulative effect.

On the aesthetic side of green architecture or sustainable design is the philosophy of designing a building that is in harmony with the natural features and resources surrounding the site. There are several key steps in designing sustainable buildings: specify 'green' building materials from local sources, reduce loads, optimize systems, and generate on-site renewable energy.

Life Cycle Assessment

A life cycle assessment (LCA) can help avoid a narrow outlook on environmental, social and economic concerns by assessing a full range of impacts associated with all cradle-to-grave stages of a process: from extraction of raw materials through materials processing, manufacture, distribution, use, repair and maintenance, and disposal or recycling. Impacts taken into account include (among others) embodied energy, global warming potential, resource use, air pollution, water pollution, and waste.

In terms of green building, the last few years have seen a shift away from a *prescriptive* approach, which assumes that certain prescribed practices are better for the environment, toward the scientific evaluation of actual performance through LCA.

Although LCA is widely recognized as the best way to evaluate the environmental impacts of buildings (ISO 14040 provides a recognized LCA methodology), it is not yet a consistent requirement of green building rating systems and codes, despite the fact that embodied energy and other life cycle impacts are critical to the design of environmentally responsible buildings.

In North America, LCA is rewarded to some extent in the Green Globes® rating system, and is part of the new American National Standard based on Green Globes, *ANSI/GBI 01-2010: Green Building Protocol for Commercial Buildings*. LCA is also included as a pilot credit in the LEED system, though a decision has not been made as to whether it will be incorporated fully into the next major revision. The state of California also included LCA as a voluntary measure in its 2010 draft *Green Building Standards Code*.

Although LCA is often perceived as overly complex and time consuming for regular use by design professionals, research organizations such as BRE in the UK and the Athena Sustainable Materials Institute in North America are working to make it more accessible.

In the UK, the BRE *Green Guide to Specifications* offers ratings for 1,500 building materials based on LCA.

In North America, the ATHENA® *EcoCalculator for Assemblies* provides LCA results for several hundred common building assembles based on data generated by its more complex parent software, the ATHENA® *Impact Estimator for Buildings*. (The EcoCalculator is available free at www.athenasmi.org.) Athena software tools are especially useful early in the design process when material choices have far-reaching implications for overall environmental impact. They allow designers to experiment with different material mixes to achieve the most effective combination.

Siting and Structure Design Efficiency

The foundation of any construction project is rooted in the concept and design stages. The concept stage, in fact, is one of the major steps in a project life cycle, as it has the largest impact on cost and performance. In designing environmentally optimal buildings, the objective is to minimize the total environmental impact associated with all life-cycle stages of the building project.

Exterior Light Shelves - Green Office Building, Denver Colorado

However, building as a process is not as streamlined as an industrial process, and varies from one building to the other, never repeating itself identically. In addition, buildings are much more complex products, composed of a multitude of materials and components each constituting various design variables to be decided at the design stage. A variation of every design variable may affect the environment during all the building's relevant life-cycle stages.

Energy Efficiency

An eco-house at Findhorn Ecovillage with a turf roof and solar panels

Green buildings often include measures to reduce energy consumption – both the embodied energy required to extract, process, transport and install building materials and operating energy to provide services such as heating and power for equipment.

As high-performance buildings use less operating energy, embodied energy has assumed much greater importance – and may make up as much as 30% of the overall life cycle energy consumption. Studies such as the U.S. LCI Database Project show buildings built primarily with wood will have a lower embodied energy than those built primarily with brick, concrete, or steel.

To reduce operating energy use, designers use details that reduce air leakage through the building envelope (the barrier between conditioned and unconditioned space). They also specify high-performance windows and extra insulation in walls, ceilings, and floors. Another strategy, passive solar building design, is often implemented in low-energy homes. Designers orient windows and walls and place awnings, porches, and trees to shade windows and roofs during the summer while maximizing solar gain in the winter. In addition, effective window placement (daylighting) can provide more natural light and lessen the need for electric lighting during the day. Solar water heating further reduces energy costs.

Onsite generation of renewable energy through solar power, wind power, hydro power, or biomass can significantly reduce the environmental impact of the building. Power generation is generally the most expensive feature to add to a building.

Water Efficiency

Reducing water consumption and protecting water quality are key objectives in sustainable building. One critical issue of water consumption is that in many areas, the demands on the supplying aquifer exceed its ability to replenish itself. To the maximum extent feasible, facilities should increase their dependence on water that is collected, used, purified, and reused on-site. The protection and conservation of water throughout the life of a building may be accomplished by designing for dual plumbing that recycles water in toilet flushing or by using water for washing of the cars. Waste-water may be minimized by utilizing water conserving fixtures such as ultra-low flush toilets and low-flow shower heads. Bidets help eliminate the use of toilet paper, reducing sewer traffic and increasing possibilities of re-using water on-site. Point of use water treatment and heating improves both water quality and energy efficiency while reducing the amount of water in circulation. The use of non-sewage and greywater for on-site use such as site-irrigation will minimize demands on the local aquifer.

Large commercial buildings with water and energy efficiency can qualify for an LEED Certification. Philadelphia's Comcast Center is the tallest building in Philadelphia. It's also one of the tallest buildings in the USA that is LEED Certified. Their environmental engineering consists of a hybrid central chilled water system which cools floor-by-floor with steam instead of water. Burn's Mechanical set-up the entire renovation of the 58 story, 1.4 million square foot sky scraper.

Materials Efficiency

Building materials typically considered to be 'green' include lumber from forests that have been certified to a third-party forest standard, rapidly renewable plant materials like bamboo and straw, dimension stone, recycled stone, recycled metal and other products that are non-toxic, reusable, renewable, and/or recyclable. For concrete a high performance or Romanself-healing concrete is available. The EPA (Environmental Protection Agency) also suggests using recycled industrial goods, such as coal combustion products, foundry sand, and demolition debris in construction projects. Energy efficient building materials and appliances are promoted in the United States through energy rebate programs.

Indoor Environmental Quality Enhancement

The Indoor Environmental Quality (IEQ) category in LEED standards, one of the five environmental categories, was created to provide comfort, well-being, and productivity of occupants. The LEED IEQ category addresses design and construction guidelines especially: indoor air quality (IAQ), thermal quality, and lighting quality.

Indoor Air Quality seeks to reduce volatile organic compounds, or VOCs, and other air impurities such as microbial contaminants. Buildings rely on a properly designed ventilation system (passively/naturally or mechanically powered) to provide adequate ventilation of cleaner air from outdoors or recirculated, filtered air as well as isolated operations (kitchens, dry cleaners, etc.) from other occupancies. During the design and construction process choosing construction materials and interior finish products with zero or low VOC emissions will improve IAQ. Most building materials and cleaning/maintenance products emit gases, some of them toxic, such as many VOCs including formaldehyde. These gases can have a detrimental impact on occupants' health, comfort, and productivity. Avoiding these products will increase a building's IEQ. LEED, HQE and Green Star contain specifications on use of low-emitting interior. Draft LEED 2012 is about to expand the scope of the involved products. BREEAM limits formaldehyde emissions, no other VOCs. MAS Certified Green is a registered trademark to delineate low VOC-emitting products in the marketplace. The MAS Certified Green Program ensures that any potentially hazardous chemicals released from manufactured products have been thoroughly tested and meet rigorous standards established by independent toxicologists to address recognized long term health concerns. These IAQ standards have been adopted by and incorporated into the following programs: (1) The United States Green Building Council (USGBC) in their LEED rating system (2) The California Department of Public Health (CDPH) in their section 01350 standards (3) The Collaborative for High Performance Schools (CHPS) in their Best Practices Manual and (4) The Business and Institutional Furniture Manufacturers Association (BIFMA) in their level® sustainability standard.

Also important to indoor air quality is the control of moisture accumulation (dampness) leading to mold growth and the presence of bacteria and viruses as well as dust mites and other organisms and microbiological concerns. Water intrusion through a building's envelope or water condensing on cold surfaces on the building's interior can enhance and sustain microbial growth. A well-insulated and tightly sealed envelope will reduce moisture problems but adequate ventilation is also necessary to eliminate moisture from sources indoors including human metabolic processes, cooking, bathing, cleaning, and other activities.

Personal temperature and airflow control over the HVAC system coupled with a properly designed building envelope will also aid in increasing a building's thermal quality. Creating a high performance luminous environment through the careful integration of daylight and electrical light sources will improve on the lighting quality and energy performance of a structure.

Solid wood products, particularly flooring, are often specified in environments where occupants are known to have allergies to dust or other particulates. Wood itself is considered to be hypo-allergenic and its smooth surfaces prevent the buildup of particles common in soft finishes like carpet. The Asthma and Allergy Foundation of American recommends hardwood, vinyl, linoleum tile or slate flooring instead of carpet. The use of wood products can also improve air quality by absorbing or releasing moisture in the air to moderate humidity.

Interactions among all the indoor components and the occupants together form the processes that determine the indoor air quality. Extensive investigation of such processes is the subject of indoor air scientific research and is well documented in the journal Indoor Air.

Operations and Maintenance Optimization

No matter how sustainable a building may have been in its design and construction, it can only remain so if it is operated responsibly and maintained properly. Ensuring operations and maintenance(O&M) personnel are part of the project's planning and development process will help retain the green criteria designed at the onset of the project. Every aspect of green building is integrated into the O&M phase of a building's life. The addition of new green technologies also falls on the O&M staff. Although the goal of waste reduction may be applied during the design, construction and demolition phases of a building's life-cycle, it is in the O&M phase that green practices such as recycling and air quality enhancement take place. O&M staff should aim to establish best practices in energy efficiency, resource conservation, ecologically sensitive products and other sustainable practices. Education of building operators and occupants is key to effective implementation of sustainable strategies in O&M services.

Waste Reduction

Green architecture also seeks to reduce waste of energy, water and materials used during construction. For example, in California nearly 60% of the state's waste comes from commercial buildings During the construction phase, one goal should be to reduce the amount of material going to landfills. Well-designed buildings also help reduce the amount of waste generated by the occupants as well, by providing on-site solutions such as compost bins to reduce matter going to landfills.

To reduce the amount of wood that goes to landfill, Neutral Alliance (a coalition of government, NGOs and the forest industry) created the website dontwastewood.com. The site includes a variety of resources for regulators, municipalities, developers, contractors, owner/operators and individuals/homeowners looking for information on wood recycling.

When buildings reach the end of their useful life, they are typically demolished and hauled to landfills. Deconstruction is a method of harvesting what is commonly considered "waste" and reclaiming it into useful building material. Extending the useful life of a structure also reduces waste – building materials such as wood that are light and easy to work with make renovations easier.

To reduce the impact on wells or water treatment plants, several options exist. "Greywater", wastewater from sources such as dishwashing or washing machines, can be used for subsurface irrigation, or if treated, for non-potable purposes, e.g., to flush toilets and wash cars. Rainwater collectors are used for similar purposes.

Centralized wastewater treatment systems can be costly and use a lot of energy. An alternative to this process is converting waste and wastewater into fertilizer, which avoids these costs and shows other benefits. By collecting human waste at the source and running it to a semi-centralized biogas plant with other biological waste, liquid fertilizer can be produced. This concept was demonstrated by a settlement in Lubeck Germany in the late 1990s. Practices like these provide soil with organic nutrients and create carbon sinks that remove carbon dioxide from the atmosphere, offsetting greenhouse gas emission. Producing artificial fertilizer is also more costly in energy than this process.

Cost and Payoff

The most criticized issue about constructing environmentally friendly buildings is the price. Photo-voltaics, new appliances, and modern technologies tend to cost more money. Most green buildings cost a premium of <2%, but yield 10 times as much over the entire life of the building. In regards to the financial benefits of green building, "Over 20 years, the financial payback typically exceeds the additional cost of greening by a factor of 4-6 times. And broader benefits, such as reductions in greenhouse gases

(GHGs) and other pollutants have large positive impacts on surrounding communities and on the planet." The stigma is between the knowledge of up-front cost vs. life-cycle cost. The savings in money come from more efficient use of utilities which result in decreased energy bills. It is projected that different sectors could save $130 Billion on energy bills. Also, higher worker or student productivity can be factored into savings and cost deductions.

Numerous studies have shown the measurable benefit of green building initiatives on worker productivity. In general it has been found that, "there is a direct correlation between increased productivity and employees who love being in their work space." Specifically, worker productivity can be significantly impacted by certain aspects of green building design such as improved lighting, reduction of pollutants, advanced ventilation systems and the use of non-toxic building materials. In "The Business Case for Green Building", the U.S. Green Building Council gives another specific example of how commercial energy retrofits increase worker health and thus productivity, "People in the U.S. spend about 90% of their time indoors. EPA studies indicate indoor levels of pollutants may be up to ten times higher than outdoor levels. LEED-certified buildings are designed to have healthier, cleaner indoor environmental quality, which means health benefits for occupants."

Studies have shown over a 20-year life period, some green buildings have yielded $53 to $71 per square foot back on investment. Confirming the rentability of green building investments, further studies of the commercial real estate market have found that LEED and Energy Star certified buildings achieve significantly higher rents, sale prices and occupancy rates as well as lower capitalization rates potentially reflecting lower investment risk.

Regulation and Operation

As a result of the increased interest in green building concepts and practices, a number of organizations have developed standards, codes and rating systems that let government regulators, building professionals and consumers embrace green building with confidence. In some cases, codes are written so local governments can adopt them as bylaws to reduce the local environmental impact of buildings.

Green building rating systems such as BREEAM (United Kingdom), LEED (United States and Canada), DGNB (Germany), CASBEE (Japan), and VERDE^{GBCe} (Spain) help consumers determine a structure's level of environmental performance. They award credits for optional building features that support green design in categories such as location and maintenance of building site, conservation of water, energy, and building materials, and occupant comfort and health. The number of credits generally determines the level of achievement.

Green building codes and standards, such as the International Code Council's draft International Green Construction Code, are sets of rules created by standards development organizations that establish minimum requirements for elements of green building such as materials or heating and cooling.

Some of the major building environmental assessment tools currently in use include:

- United States: International Green Construction Code (IGCC)

International Frameworks and Assessment Tools

IPCC Fourth Assessment Report

Climate Change 2007, the Fourth Assessment Report (AR4) of the United Nations Intergovernmental Panel on Climate Change (IPCC), is the fourth in a series of such reports. The IPCC was established by the World Meteorological Organization (WMO) and the United Nations Environment Programme (UNEP) to assess scientific, technical and socio-economic information concerning climate change, its potential effects and options for adaptation and mitigation.

UNEP and Climate Change

United Nations Environment Program UNEP works to facilitate the transition to low-carbon societies, support climate proofing efforts, improve understanding of climate change science, and raise public awareness about this global challenge.

GHG Indicator

The Greenhouse Gas Indicator: UNEP Guidelines for Calculating Greenhouse Gas Emissions for Businesses and Non-Commercial Organizations

Agenda 21

Agenda 21 is a programme run by the United Nations (UN) related to sustainable development. It is a comprehensive blueprint of action to be taken globally, nationally and locally by organizations of the UN, governments, and major groups in every area in which humans impact on the environment. The number 21 refers to the 21st century.

Fidic's PSM

The International Federation of Consulting Engineers (FIDIC) Project Sustainability Management Guidelines were created in order to assist project engineers and other stakeholders in setting sustainable development goals for their projects that are recognized and accepted by as being in the interests of society as a whole. The process is also intended to allow the alignment of project goals with local conditions and priorities and to assist those involved in managing projects to measure and verify their progress.

The Project Sustainability Management Guidelines are structured with Themes and Sub-Themes under the three main sustainability headings of Social, Environmental and Economic. For each individual Sub-Theme a core project indicator is defined along with guidance as to the relevance of that issue in the context of an individual project.

The Sustainability Reporting Framework provides guidance for organizations to use as the basis for disclosure about their sustainability performance, and also provides stakeholders a universally applicable, comparable framework in which to understand disclosed information.

The Reporting Framework contains the core product of the Sustainability Reporting Guidelines, as well as Protocols and Sector Supplements. The Guidelines are used as the basis for all reporting. They are the foundation upon which all other reporting guidance is based, and outline core content for reporting that is broadly relevant to all organizations regardless of size, sector, or location. The Guidelines contain principles and guidance as well as standard disclosures – including indicators – to outline a disclosure framework that organizations can voluntarily, flexibly, and incrementally, adopt.

Protocols underpin each indicator in the Guidelines and include definitions for key terms in the indicator, compilation methodologies, intended scope of the indicator, and other technical references.

Sector Supplements respond to the limits of a one-size-fits-all approach. Sector Supplements complement the use of the core Guidelines by capturing the unique set of sustainability issues faced by different sectors such as mining, automotive, banking, public agencies and others.

IPD Environment Code

The IPD Environment Code was launched in February 2008. The Code is intended as a good practice global standard for measuring the environmental performance of corporate buildings. Its aim is to accurately measure and manage the environmental impacts of corporate buildings and enable property executives to generate high quality, comparable performance information about their buildings anywhere in the world. The Code covers a wide range of building types (from offices to airports) and aims to inform and support the following;

- Creating an environmental strategy
- Inputting to real estate strategy
- Communicating a commitment to environmental improvement
- Creating performance targets
- Environmental improvement plans

- Performance assessment and measurement
- Life cycle assessments
- Acquisition and disposal of buildings
- Supplier management
- Information systems and data population
- Compliance with regulations
- Team and personal objectives

IPD estimate that it will take approximately three years to gather significant data to develop a robust set of baseline data that could be used across a typical corporate estate.

ISO 21931

ISO/TS 21931:2006, Sustainability in building construction—Framework for methods of assessment for environmental performance of construction works—Part 1: Buildings, is intended to provide a general framework for improving the quality and comparability of methods for assessing the environmental performance of buildings. It identifies and describes issues to be taken into account when using methods for the assessment of environmental performance for new or existing building properties in the design, construction, operation, refurbishment and deconstruction stages. It is not an assessment system in itself but is intended be used in conjunction with, and following the principles set out in, the ISO 14000 series of standards.

Groundwater Banking

Groundwater banking is a water management tool designed to increase water supply reliability. Groundwater can be created by using dewatered aquifer space to store water during the years when there is abundant rainfall. It can then be pumped and used during years that don't have a surplus of water. People can manage the use of groundwater to benefit society through the purchasing and selling of these groundwater rights. The surface water should be used first, and then the groundwater will be used when there isn't enough surface water to meet demands. The groundwater will reduce the risk of relying on surface water and will maximize expected income. There are regulatory storage-type aquifer recovery and storage systems which when water is injected into it gives the right to withdraw the water later on. Groundwater banking has been implemented into semiarid and arid southwestern United States because this is where there is the most need for extra water. The overall goal is to transfer water from low-value to high-value uses by bringing buyers and sellers together.

Groundwater Storage Concepts

The bank is an aquifer used as an underground storage tank, and the recharge of water causes an increase in the volume of water stored in the aquifer to have increasing water levels. In the case of a withdrawal there would be a decrease in water levels. The amount of water depends also on a couple of other factors including groundwater pumping by other users, leakage, and natural recharge. The recharge of water by land application or injection increases the volume of water, and then some of the water will be used at a future time. It can be looked at as inputs of water minus outputs is equal to the change in water storage.

Another aspect is the hydrology which is the difference between dynamic and static response to recharge and abstraction. The water levels will rise or fall in the well during recharge and recovery. Once recharge and recovery stops the water levels return to background levels, and one of the main issues is the change in static water levels after the dynamic response from recharge or recovery disappears.

There can be some technical issues with the aquifer response to manage recharge and recovery. If the aquifer is hydraulically connected to body of water on the surface there are increases in the water table elevation as the result of managed recharge. This could increase the rate of discharge or decrease the amount of induced recharge, and both of these cause water to leave the basin. The recharge and recovery could also affect the lateral and vertical groundwater flow into the aquifer. There isn't always a one-to-one correspondence between the volume of water and the change in storage.

Methods of Groundwater Banking

'"Groundwater banking"' is accomplished in two ways: through in-lieu and direct recharge. In-lieu recharge is storing water by utilizing surface water "in-lieu" of pumping groundwater, thereby storing an equal amount in the groundwater basin. In-lieu recharge is the renewable surface water used to irrigate the farmland in place of using regular groundwater. This is helping to save more groundwater because the water stays in the aquifer to be used later. Direct recharge is storing water by allowing it to percolate directly to storage in the groundwater basin. With direct recharge it floods an area so that water seeps through the ground to get to the aquifers. The water is then pumped out when there is more of a demand with the use of recovery wells.

Pros and Cons

There are some disadvantages to retrieving this groundwater. As groundwater is withdrawn from below the Earth's surface, the ground above settles. This settling of land, known as subsidence, can fracture roads and building foundations and can burst water, sewer, and gas lines. This method hasn't been well tested yet, so there could be some negative impacts on the environment. Water is a public resource and this could make water become a private industry.

Groundwater banking can be compared and contrasted to the use of surface reservoirs. Groundwater banking has many advantages over the use of surface reservoirs. The projects do not cost as much to construct and store the same amount, if not more, than the surface reservoirs. The bank will have less of an impact on the environment than a surface reservoir. The water that is in the bank will no longer be exposed to evaporation, but several feet per year of water is lost in the reservoirs. It is more reliable to use when the climate is changing and can respond to seasonal changes better to manage the water than the surface reservoirs.

There are also some disadvantages to groundwater banking over surface reservoirs. There are energy costs to recovering the water and these costs are usually more than the reservoirs. There is also a pumping capacity and when the demands change during the year the productiveness can be limited.

Not all groundwater is used when sold. Some groundwater is being studied for its benefits. Groundwater banking and aquifer storage systems are being explored to control flooding during times of high precipitation.

The groundwater is being traded in many regions. There are trades even in the United States. The city of San Antonio, Texas is the largest city in the United States that relies solely on groundwater for its municipal supply.

Feasibility of Groundwater Recharge on Agricultural Land

There haven't been many successful trials of groundwater banking on agricultural land since the land is usually privately owned. The owners have to be on board with the practice of groundwater banking knowing what the risks and best practices entail. A study was done to find a Soil Agricultural Groundwater Banking Index (SAGBI) which evaluates soil suitability for the use of groundwater banking in California. There are five factors that determine the feasibility of groundwater recharge on agricultural land: deep percolation, root zone residence time, topography, chemical limitations, and soil surface conditions. The five factors were modeled using United States Department of Agriculture Natural Resources Conservation Service (USDA-NRCS) digital soil survey data.

For the deep percolation factor a high rate of water transmission through the soil profile and into the aquifer below is the key to successful groundwater banking. It becomes more important when there is flooding since it could be used as the main water source. It is derived from the saturated hydraulic conductivity of the limiting layer. Saturated hydraulic conductivity measures soil permeability when the soil is saturated.

When looking at root zone residence time factor it was found that a prolonged duration of saturated conditions in the root zone has the possibility to cause damage to perennial crops. If the soil causes a bud break it is more likely that the crop will become damaged. Most crops aren't able to withstand long periods of saturated conditions in the root zone. The root zone residence time estimates the likelihood of having good enough drainage within the root zone once water is applied.

The topography of land for spreading water across fields has the best outcome when there is level topography. Level topography works the best because it holds water better on the landscape which allows infiltration across large areas. Infiltration reduces ponding and minimizes erosion by runoff.

The chemical limitations factor is related to the salinity which is a threat to the sustainability of agriculture and groundwater. This factor was determined by electrical conductivity (EC) of the soil which measures the soil salinity. The best soil has the lowest levels of salinity. Soil also has pesticides and nitrate, but it is unable to be evaluated due to the dependency on management history.

The surface condition factor is when banking by flood spreading can change the soil surfaces physical conditions. Infiltrations limits can be caused by quality and depth of water that could lead to the destruction of aggregates, the formation of physical crusts, and compaction. To determine soil condition two factors were examined: soil erosion factor and sodium absorption ratio (SAR).

To determine the feasibility of groundwater banking each of the five factors were assigned a weight to how significant it was, and then a SAGBI score was calculated. The weights were 27.5% deep percolation, 27.5% root zone residence time, 20% topography, 20% chemical limitations, and 5% surface conditions. Of the 17.5 million acres of agricultural land examined only 5 million acres were considered soils with excellent, good, and moderately good suitability.

Agricultural groundwater banking can be associated with financial risk which may cause crop loss, so in the end the loss may exceed the benefits of water saving. Adoption of this practice would require support to protect growers from risk of crop failure.

Groundwater Accounting System

The accounting system tracks the recharge and withdrawals of stored water and it can include a market system to reserve the storage of water. Depositors could earn credits for the recharge of water which can be used later on for the water recovered from the bank. The banking system could then be set up to allow trading of credits. There are several objectives of the water bank accounting system: track water deposits, withdrawals, and to control the amount, timing, and location of withdrawals by the participants. If groundwater isn't regulated there is more of a chance for freeriding and overuse. There needs to be sustainability of the system in order to continue with the operation of the system or there would be no point to using it.

The accounting method that will be used is the double-entry accounting method, so every transaction is recorded as a debit and a credit in separate ledger accounts. This also allows for tracking of inventory in asset accounts and claims to inventory in ownership accounts. Deposits happen when more water is stored in an aquifer than there is supposed to be. The recharge of water by a member would be a credit in a member's account and a liability in the bank's account. For the bank to be successful then both ledgers have to be balanced, so the right to water in a member's accounts should be equal to the amount of water that can be recovered from the system. If the right to water is greater than liabilities then the bank is insolvent, and this will become a problem when a drought occurs.

There are some issues that could arise from using this water bank accounting system. These problems were evaluated for the Las Posas Basin groundwater bank and Fox Canyon Groundwater Management Agency (FCGMA) has jurisdiction over the project. FCGMA reported that the accumulation of credits has been increasing for banks. What can happen is the accumulated credits can become greater than the annual abstraction rate. The volume of credits accumulating exceed the amount of water that can be taken out during a short-time period. This will cause a threat to the regional groundwater resource or even depressions in groundwater elevations, land subsidence, and seawater intrusion.

Regulatory Framework

The banking systems need regulatory control over the basin to implement the withdrawal rates and to ensure that other participants won't abstract too much stored water. The best scenario would be that the bank owner or participants would be the main users to ensure that abstractions are controlled. If this is not the case, then there must be another way to control the number and amount of abstractions happening. It needs to be clear who has priority over stored water, so that when abstractions are constrained it is known who will get the water first. There can be problems when multiple entities have jurisdiction over a project, and this can cause regulatory and organizational challenges. There are some generally accepted rules and also many of the issues are handled through state-specific concepts. One of the main requirements is an action needs to be in place so that the stored water isn't being abstracted by other users who aren't involved in the system. These frameworks rely on the knowledge of hydrogeology to determine the success of a system, and the systems need to provide benefits to prove it was worth building.

Economics

The different projects can become economically efficient by maximizing the benefits of the limited resource (water). To maximize efficiency the users need to find where marginal cost is equal to marginal benefit. It is important for supply to equal demand like in the figure below. The use of water becomes a negative externality when there is rivalry and the property rights are not well-defined. The way to eliminate some of the negative externality is there can be a tax placed on the resource to increase the marginal cost. When they tax the right amount the user will use the resource at the socially acceptable level. The other way to affect the externality is to create a subsidy. A subsidy will increase the marginal benefit in order to get to the socially accepted level of use for the resource.

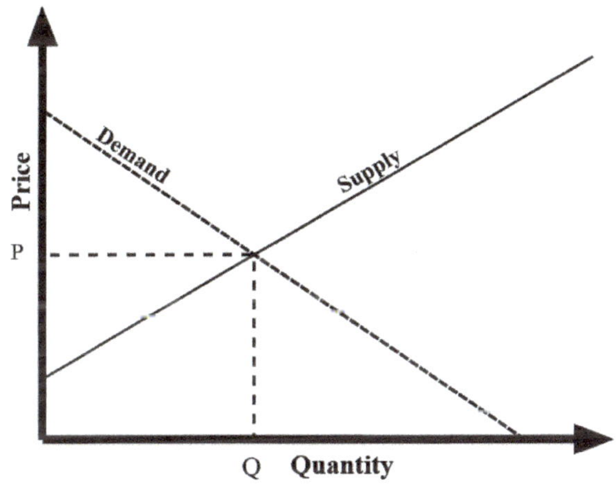

Water is not a homogeneous commodity for several reasons which include sensitivity to location, time of use, form of the water, and administrative responses. The use of groundwater banking can make water a more homogenous commodity. This can create a market value which will enhance private investment increasing the benefits. It will also align marginal benefit with marginal cost causing the market to come to a economically-efficient level.

Water has high transaction costs and create market barriers which devalues the use to society restricting the reallocation of resources. The demand does not change when there is a market barrier so there will be many unpleasant people if they don't get their share of water. If a bank is in the process of being made the removal of market-access barriers can be part of the negotiations, but it isn't necessarily the bank itself that is the cause. Groundwater banking could reduce transaction costs because each individual won't have to analyze each transaction.

Permeable Paving

Permeable paving is a range of sustainable materials and techniques for permeable pavements with a base and subbase that allow the movement of stormwater through the surface. In addition to reducing runoff, this effectively traps suspended solids and filters pollutants from the water. Examples include roads, paths, lawns and lots that are subject to light vehicular traffic, such as car/parking lots, cycle-paths, service or emergency access lanes, road and airport shoulders, and residential sidewalks and driveways.

Permeable paving demonstration

Stone paving in Santarém, Portugal

Although some porous paving materials appear nearly indistinguishable from nonporous materials, their environmental effects are qualitatively different. Whether it is pervious concrete, porous asphalt, paving stones or concrete or plastic-based pavers, all these pervious materials allow stormwater to percolate and infiltrate the surface areas, traditionally impervious to the soil below. The goal is to control stormwater at the source, reduce runoff and improve water quality by filtering pollutants in the substrata layers.

Description and Applications

Permeable solutions can be based on: porous asphalt and concrete surfaces, concrete pavers (permeable interlocking concrete paving systems - PICP), or polymer-based grass pavers, grids and geocells. Porous pavements and concrete pavers (actually the voids in-between them) enable stormwater to drain through a stone base layer for on-site infiltration and filtering. Polymer based grass grid or cellular paver systems provide load bearing reinforcement for unpaved surfaces of gravel or turf.

Grass pavers, plastic turf reinforcing grids (PTRG), and geocells (cellular confinement systems) are honeycombed 3D grid-cellular systems, made of thin-walled HDPE plastic or other polymer alloys. These provide grass reinforcement, ground stabilization and gravel retention. The 3D structure reinforces infill and transfers vertical loads from the surface, distributing them over a wider area. Selection of the type of cellular grid depends to an extent on the surface material, traffic and loads. The cellular grids are installed on a prepared base layer of open-graded stone (higher void spacing) or engineered stone (stronger). The surface layer may be compacted gravel or topsoil seeded with grass and fertilizer. In addition to load support, the cellular grid reduces compaction of the soil to maintain permeability, while the roots improve permeability due to their root channels.

In new suburban growth, porous pavements protect watersheds. In existing built-up areas and towns, redevelopment and reconstruction are opportunities to implement stormwater water management practices. Permeable paving is an important component in Low Impact Development (LID), a process for land development in the United States that attempts to minimize impacts on water quality and the similar concept of sustainable drainage systems (SuDS) in the United Kingdom.

The infiltration capacity of the native soil is a key design consideration for determining the depth of base rock for stormwater storage or for whether an underdrain system is needed.

Advantages

Managing Runoff

Permeable paving surfaces have been demonstrated as effective in managing runoff from paved surfaces. Large volumes of urban runoff causes serious erosion and siltation in surface water bodies.

Controlling Pollutants

Permeable paving surfaces keep the pollutants in place in the soil or other material underlying the roadway, and allow water seepage to groundwater recharge while preventing the stream erosion problems. They capture the heavy metals that fall on them, preventing them from washing downstream and accumulating inadvertently in the environment. In the void spaces, naturally occurring micro-organisms digest car oils, leaving little but carbon dioxide and water. Rainwater infiltration is usually less than that of an impervious pavement with a separate stormwater management facility somewhere downstream..in areas where infiltration is not possible due to unsuitable soil conditions permeable pavements are used in the attenuation mode where water is retained in the pavement and slowly released to surface water systems between storm events.

Trees

Permeable pavements may give urban trees the rooting space they need to grow to full size. A "structural-soil" pavement base combines structural aggregate with soil; a porous surface admits vital air and water to the rooting zone. This integrates healthy ecology and thriving cities, with the living tree canopy above, the city's traffic on the ground, and living tree roots below. The benefits of permeables on urban tree growth have not been conclusively demonstrated and many researchers have observed tree growth is not increased if construction practices compact materials before permeable pavements are installed.

Disadvantages

Runoff Volumes

Permeable pavements are designed to replace Effective Impervious Areas (EIAs), not to manage stormwater from other impervious surfaces on site. Use of this technique must be part of an overall on site management system for stormwater, and is not a replacement for other techniques.

Also, in a large storm event, the water table below the porous pavement can rise to a higher level preventing the precipitation from being absorbed into the ground. The additional water is stored in the open graded crushed drain rock base and remains until the subgrade can absorb the water. For clay-based soils, or other low to 'non'-draining soils, it is important to increase the depth of the crushed drain rock base to allow additional capacity for the water as it waits to be infiltrated.

The best way to prevent this problem is to understand the soil infiltration rate, and design the pavement and base depths to meet the volume of water. Or, allow for adequate rain water run off at the pavement design stage.

Pollutant Load

Highly contaminated runoff can be generated by some land uses where pollutant concentrations exceed those typically found in stormwater. These "hot spots" include commercial nurseries, recycling facilities, fueling stations, industrial storage, marinas, some outdoor loading facilities, public works yards, hazardous materials generators (if containers are exposed to rainfall), vehicle service and maintenance areas, and vehicle and equipment washing and steam cleaning facilities. Since porous pavement is an infiltration practice, it should not be applied at stormwater hot spots due to the potential for groundwater contamination. All contaminated runoff should be prevented from entering municipal storm drain systems by using best management practices (BMPs) for the specific industry or activity.

Weight and Traffic Volumes

Reference sources differ on whether low or medium traffic volumes and weights are appropriate for porous pavements. For example, around truck loading docks and areas of high commercial traffic, porous pavement is sometimes cited as being inappropriate. However, given the variability of products available, the growing number of existing installations in North America and targeted research by both manufacturers and user agencies, the range of accepted applications seems to be expanding. Some concrete paver companies have developed products specifically for industrial applications. Working examples exist at fire halls, busy retail complex parking lots, and on public and private roads, including intersections in parts of North America with quite severe winter conditions.

Siting

Permeable pavements may not be appropriate when land surrounding or draining into the pavement exceeds a 20 percent slope, where pavement is down slope from buildings or where foundations have piped drainage at their footers. The key is to ensure that drainage from other parts of a site is intercepted and dealt with separately rather than being directed onto permeable surfaces.

Climate

Cold climates may present special challenges. Road salt contains chlorides that could migrate through the porous pavement into groundwater. Snow plow blades could catch block edges and damage surfaces. Sand cannot be used for snow and ice control on perveous asphalt or concrete because it will plug the pores and reduce permeability. Infiltrating runoff may freeze below the pavement, causing frost heave, though design modifications can reduce this risk. These potential problems do not mean that porous pavement cannot be used in cold climates. Porous pavement designed to reduce frost heave has been used successfully in Norway. Furthermore, experience suggests that rapid drainage below porous surfaces increases the rate of snow melt above.

Cost

Some estimates put the cost of permeable paving at two to three times that of conventional asphalt paving. Using permeable paving, however, can reduce the cost of providing larger or more stormwater BMPs on site, and these savings should be factored into any cost analysis. In addition, the off-site environmental impact costs of not reducing on-site stormwater volumes and pollution have historically been ignored or assigned to other groups (local government parks, public works and environmental restoration budgets, fisheries losses, etc.) The City of Olympia, Washington is studying the use of pervious concrete quite closely and finding that new stormwater regulations are making it a viable alternative to storm water.

Longevity and Maintenance

Some permeable pavements require frequent maintenance because grit or gravel can block the open pores. This is commonly done by industrial vacuums that suck up all the sediment. If maintenance is not carried out on a regular basis, the porous pavements can begin to function more like impervious surfaces. With more advanced paving systems the levels of maintenance needed can be greatly decreased, elastomerically bound glass pavements requires less maintenance than regular concrete paving as the glass bound pavement has 50% more void space.

Plastic grid systems, if selected and installed correctly, are becoming more and more popular with local government maintenance personnel owing to the reduction in maintenance efforts: reduced gravel migration and weed suppression in public park settings.

Some permeable paving products are prone to damage from misuse, such as drivers who tear up patches of plastic & gravel grid systems by "joy riding" on remote parking lots at night. The damage is not difficult to repair but can look unsightly in the meantime. Grass pavers require supplemental watering in the first year to establish the vegetation, otherwise they may need to be re-seeded. Regional climate also means that most grass applications will go dormant during the dry season. While brown vegetation is only a matter of aesthetics, it can influence public support for this type of permeable paving.

Traditional permeable concrete paving bricks tend to lose their color in relatively short time which can be costly to replace or clean and is mainly due to the problem of efflorescence.

Efflorescence

Efflorescence is a hardened crystalline deposit of salts, which migrate from the center of concrete or masonry pavers to the surface to form insoluble calcium carbonates that harden on the surface. Given time, these deposits form much like how a stalactite takes shape in a cave, except in this case on a flat surface. Efflorescence usually appears white, gray or black depending on the region.

Over time efflorescence begins to negatively affect the overall appearance of masonry/concrete and may cause the surfaces to become slippery when exposed to moisture. If left unchecked, this efflorescence will harden whereby the calcium/lime deposits begin to affect the integrity of the cementatious surface by slowly eroding away the cement paste and aggregate. In some cases it will also discolor stained or coated surfaces.

Efflorescence forms more quickly in areas that are exposed to excessive amounts of moisture such as near pool decks, spas, and fountains or where irrigation runoff is present. As a result, these affected regions become very slick when wet thereby causing a significant loss of "friction coefficient". This can be of serious concern especially as a public safety issue to individuals, principals and property owners by exposing them to possible injury and increased general liability claims.

Efflorescence remover chemicals can be used to remove calcium/lime build-up without damaging the integrity of the paving surface.

Types

Installation of porous pavements is no more difficult than that of dense pavements, but has different specifications and procedures which must be strictly adhered to. Nine different families of porous paving materials present distinctive advantages and disadvantages for specific applications. Here are examples:

Pervious Concrete

Pervious concrete is widely available, can bear frequent traffic, and is universally accessible. Pervious concrete quality depends on the installer's knowledge and experience.

Plastic Grids

Plastic grids allow for a 100% porous system using structural grid systems for containing and stabilizing either gravel or turf. These grids come in a variety of shapes and sizes depending on use; from pathways to commercial parking lots. These systems have been used readily in Europe for over a decade, but are gaining popularity in North America due to requirements by government for many projects to meet LEED environmental building standards. Plastic grid system are also popular with homeowners due to their lower cost to install, ease of installation, and versatility. The ideal design for this type of grid system is a closed cell system, which prevents gravel/sand/turf from migrating laterally. It is also known as Grass pavers / Turf Pavers in India

Porous Asphalt

Porous asphalt is produced and placed using the same methods as conventional asphalt concrete; it differs in that fine (small) aggregates are omitted from the asphalt mixture. The remaining large, single-sized aggregate particles leave open voids that give the material its porosity and permeability. To ensure pavement strength, fiber may be added to the mix or a polymer-modified asphalt binder may be used. Generally, porous asphalt pavements are designed with a subsurface reservoir that holds water that passes through the pavement, allowing it to evaporate and/or percolate slowly into the surround soils.

Open-graded friction courses (OGFC) are a porous asphalt surface course used on highways to improve driving safety by removing water from the surface. Unlike a full-depth porous asphalt pavement, OGFCs do not drain water to the base of a pavement. Instead, they allow water to infiltrate the top 3/4 to 1.5 inch of the pavement and then drain out to the side of the roadway. This can improve the friction characteristics of the road and reducing road spray.

Single-Sized Aggregate

Single-sized aggregate without any binder, e.g. loose gravel, stone-chippings, is another alternative. Although it can only be safely used in very low-speed, low-traffic settings, e.g. car-parks and drives, its potential cumulative area is great.

Porous Turf

Grass pavement

Porous turf, if properly constructed, can be used for occasional parking like that at churches and stadia. Plastic turf reinforcing grids can be used to support the increased load. Living turf transpires water, actively counteracting the "heat island" with what appears to be a green open lawn.

Permeable Interlocking Concrete Pavements

Permeable interlocking concrete pavements are concrete (or stone) units with open, permeable spaces between the units. They give an architectural appearance, and can bear both light and heavy traffic, particularly interlocking concrete pavers, excepting high-volume or high-speed roads. Some products are polymer-coated and have an entirely porous face.

Permeable Clay Brick Pavements

Permeable clay brick pavements are fired clay brick units with open, permeable spaces between the units. Clay pavers provide a durable surface that allows stormwater runoff to permeate through the joints.

Resin Bound Paving

Resin bound paving is a mixture of resin binder and aggregate. Clear resin is used to fully coat each aggregate particle before laying. Enough resin is used to allow each aggregate particle to adhere to one another and to the base yet leave voids for water to permeate through. Resin bound paving provides a strong and durable surface that is suitable for pedestrian and vehicular traffic in applications such as pathways, driveways, car parks and access roads.

Bound Recycled Glass Porous Pavement

Elastomerically bound recycled glass porous pavement consisting of bonding processed post consumer glass with a mixture of resins, pigments, granite and binding agents. Approximately 75 percent of glass in the U.S. is disposed in landfills.

Seawall

An example of a modern seawall in Ventnor on the Isle of Wight

People socializing and walking at the Malecón, Havana

Seawall at Urangan, Queensland

A seawall (or sea wall) is a form of coastal defense constructed where the sea, and associated coastal processes, impact directly upon the landforms of the coast. The purpose of a seawall is to protect areas of human habitation, conservation and leisure activities from the action of tides and waves. As a seawall is a static feature it will conflict with the dynamic nature of the coast and impede the exchange of sediment between land and sea.

The coast is generally a high-energy, dynamic environment with spatial variations occurring over a wide range of temporal scales. The shoreline is part of the coastal interface which is exposed to a wide range of erosional processes arising from fluvial, aeolian and terrestrial sources, meaning that a combination of denudational processes will work against a seawall. Given the natural forces to which seawalls are constantly subjected, maintenance (and eventually replacement) is an ongoing requirement if they are to provide an effective long-term solution.

The many types in use today reflect both the varying physical forces they are designed to withstand, and location specific aspects, such as: local climate, coastal position, wave regime, and value of landform. Seawalls are classified as a hard engineering shore based structure used to provide protection and to lessen coastal erosion. However, a range of environmental problems and issues may arise from the construction of a seawall, including disrupting sediment movement and transport patterns, which are discussed in more detail below. Combined with a high construction cost, this has led to an increasing use of other soft engineering coastal management options such as beach replenishment.

Seawalls may be constructed from a variety of materials, most commonly: reinforced concrete, boulders, steel, or gabions. Additional seawall construction materials may include: vinyl, wood, aluminium, fibreglass composite, and with large biodegrable sandbags made of jute and coir. In the UK, *sea wall* also refers to an earthen bank used to create a polder, or a dike construction.

Types

A seawall works by reflecting incident wave energy back into the sea, thereby reducing the energy and erosion which the coastline would otherwise be subjected to. In addition to their unsightly visual appearance, two specific weaknesses of seawalls exist. Firstly, wave reflection induced by the wall may result in scour and subsequent lowering of the sand level of the fronting beach. Secondly, seawalls may accelerate erosion of adjacent, unprotected coastal areas because they affect the littoral drift process. The design and type of seawall that is appropriate depends on aspects specific to the location, including the surrounding erosion processes. There are three main types of seawalls: vertical, curved or stepped, and mounds, as set out in the table:

<table>
<tr><th colspan="5">Seawall types</th></tr>
<tr><th>Type</th><th>Illustration</th><th>Advantages</th><th>Disadvantages</th><th>Example</th></tr>
<tr><td rowspan="2">Vertical</td><td colspan="4">Vertical seawalls are built in particularly exposed situations. These reflect wave energy. Under storm conditions a non-breaking standing wave pattern can form, resulting in a stationary clapotic wave which moves up and down but does not travel horizontally. These waves promote erosion at the toe of the wall and can cause severe damage to the sea wall. In some cases piles are placed in front of the wall to lessen wave energy slightly.</td></tr>
<tr><td>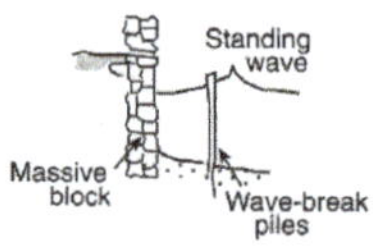
</td><td>• The first implemented, most easily designed and constructed type of seawall.
• Vertical sea walls deflect wave energy away from the coast.
• Loose rubble can absorb wave energy.</td><td>• These are partial to a lot of expensive damage in a short period of time.
• Vertical design can be undercut by high-wave energy environments over a long period of time.</td><td></td></tr>
<tr><td rowspan="2">Curved</td><td colspan="4">Curved or stepped seawalls are designed to enable waves to break to dissipate wave energy and to repel waves back to the sea. The curve can also prevent the wave overtopping the wall and provides additional protection for the toe of the wall.</td></tr>
<tr><td>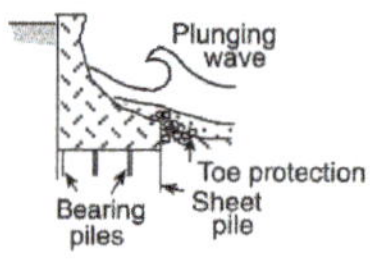
</td><td>• Concave structure introduces a dissipative element.
• The curve can prevent waves from overtopping the wall and provides extra protection for the toe of the wall
• Curved seawalls aim to re-direct most of the incident energy, resulting in low reflected waves and much reduced turbulence.</td><td>• More complex engineering and design process.
• The deflected waves can scour material at the base of the wall causing them to become undermined.</td><td></td></tr>
</table>

<table>
<tr><td rowspan="2">Mound</td><td colspan="4">Mound type seawalls, using revetments or riprap, are used in less demanding settings where lower energy erosional processes operate. The least exposed sites involve the lowest-cost bulkheads and revetments of sand bags or geotextiles. These serve to armour the shore and minimise erosion and may be either watertight or porous, which allows water to filter through after the wave energy has been dissipated.</td></tr>
<tr><td>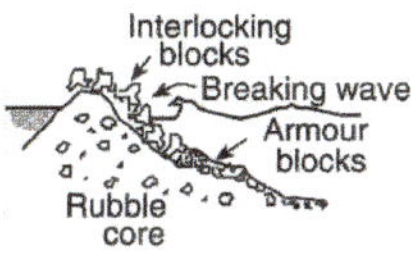
</td><td><ul><li>Current designs use porous designs of rock, concrete armour.</li><li>Slope and loose material ensure maximum dissipation of wave energy.</li><li>Lower cost option.</li></ul></td><td><ul><li>Less durable.</li><li>Shorter life expectancy.</li><li>Cannot withstand or protect from high-energy conditions effectively.</li></ul></td><td></td></tr>
</table>

Trade-Offs

A cost benefit approach is an effective way to determine whether a seawall is appropriate and whether the benefits are worth the expense. Besides controlling erosion, consideration must be given to the effects of hardening a shoreline on natural coastal ecosystems and human property or activities. A seawall is a static feature which can conflict with the dynamic nature of the coast and impede the exchange of sediment between land and sea. The table below summarises some positive and negative effects of seawalls which can be used when comparing their effectiveness with other coastal management options, such as beach nourishment.

<table>
<tr><th colspan="2">Advantages and disadvantages of seawalls according to Short (1999)</th></tr>
<tr><th>Advantages</th><th>Disadvantages</th></tr>
<tr><td><ul><li>Long term solution in comparison to soft beach nourishment.</li><li>Effectively minimizes loss of life in extreme events and damage to property caused by erosion.</li><li>Can exist longer in high energy environments in comparison to ‘soft’ engineering methods.</li><li>Can be used for recreation and sightseeing.</li><li>Forms a hard and strong coastal defense.</li></ul></td><td><ul><li>Very expensive to construct.</li><li>Can cause beaches to dissipate rendering them useless for beach goers.</li><li>Scars the very landscape that they are trying to save and provides an ‘eyesore.’</li><li>Reflected energy of waves leading to scour at base.</li><li>Can disrupt natural shoreline processes and destroy shoreline habitats such as wetlands and intertidal beaches.</li><li>Altered sediment transport processes can disrupt sand movement that can lead to increased erosion down drift from the structure.</li></ul></td></tr>
</table>

Generally seawalls can be a successful way to control coastal erosion, but only if they are constructed well and out of materials which can withstand the force of ongoing wave energy. Some understanding is needed of the coastal processes and morphodynamics specific to the seawall location. Seawalls can be very helpful; they can offer a more long-term solution than soft engineering options, additionally providing recreation opportunities and protection from extreme events as well as everyday erosion. Extreme natural events expose weaknesses in the performance of seawalls, and analyses of these can lead to future improvements and reassessment.

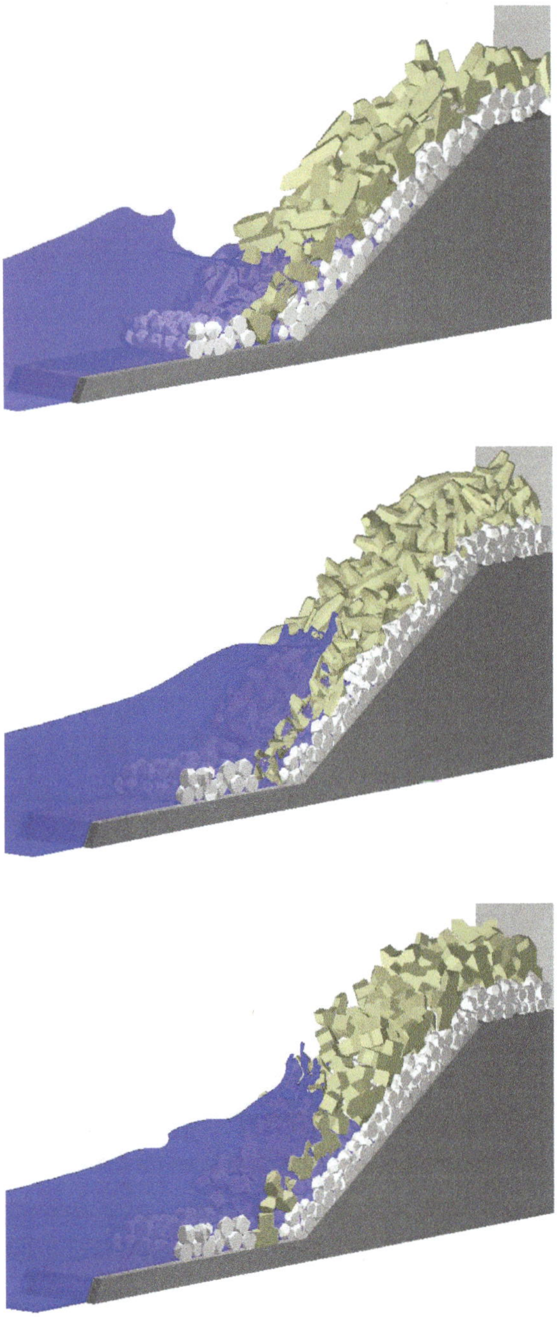

3D simulation of wave motion near a sea wall.

Issues

Sea Level Rise

Sea level rise creates an issue for seawalls worldwide as it raises both the mean normal water level and the height of waves during extreme weather events, which the current seawall heights may be unable to cope with (Allan *et al.* 1999). The most recent analyses of long, good-quality tide gauge records (corrected for GIA and when possible for other vertical land motions by the Global Positioning System, GPS) indicate a mean rate of sea level rise of 1.6–1.8 mm/yr over the twentieth century. The Intergovernmental Panel on Climate Change (IPCC) (1997) suggested that sea level rise over the next 50 – 100 years will accelerate with a projected increase in global mean sea level of +18 cm by 2050 AD. This data is reinforced by Hannah (1990) who calculated similar statistics including a rise of between +16-19.3 cm throughout 1900–1988. Super storm Sandy of 2012 is an example of the devastating effects rising sea levels can cause when mixed with a perfect storm. According to Nabil Ismail, Professor of Coastal Engineering at the Maritime Academy for Science and Technology and S. Jeffress Williams, Dept. of Geology and Geophysics at The University of Hawaii at Manoa stated that Super storm Sandy sent a storm surge of 4–5 m onto New Jersey's and New York's barrier island and urban shorelines, estimated at $70 billion in damage. EGU General Assembly Conference Abstracts. Vol. 15. 2013.This problem could be overcome by further modelling and determining the extension of height and reinforcement of current seawalls which needs to occur for safety to be ensured in both situations.

Extreme Events

Extreme events also pose a problem as it is not easy for people to predict or imagine the strength of hurricane or storm induced waves compared to normal, expected wave patterns. An extreme event can dissipate hundreds of times more energy than everyday waves, and calculating structures which will stand the force of coastal storms is difficult and, often the outcome can become unaffordable. For example, Omaha Beach seawall in New Zealand was designed to prevent erosion from everyday waves only, and when a storm in 1976 carved out 10 m behind the existing seawall the whole structure was destroyed (GeoResources, 2001).

Other Issues

Some further issues include: lack of long term trend data of seawall effects due to a relatively short duration of data records; modelling limitations and comparisons of different projects and their effects being invalid or unequal due to different beach types; materials; currents; and environments (Christchurch City Council, 2009). Lack of maintenance is also a major issue with sea walls. In a post on October 21, 2013 by NBC reporter Lucas Seiler, more than 5,000 feet of seawall is crumbling in Punta Gorda, Florida. Residents of the area pay hundreds of dollars each year into a seawall repair program. The problem is that most of the seawalls are over a half century old and are being destroyed by only heavy downpours. If not kept in check, seawalls lose effectiveness and become expensive to repair.

Historical Examples

Seawall construction has existed since ancient times. In the 1st century BCE, Romans built a seawall / breakwater at Caesarea Maritima creating an artificial harbor (Sebastos Harbor). The construction used Pozzolana concrete which hardens in contact with sea water. Barges were constructed and filled with the concrete. They were floated into position and sunk. The resulting harbor / breakwater / sea wall is still in existence today - more than 2000 years later.

More recently, sea walls were constructed in 1623 in Canvey Island, UK, when great floods of the Thames estuary occurred, prompting the construction of protection for further events in this flood prone area (Council of Europe, 1999). Since then, seawall design has become more complex and intricate in response to an improvement in materials, technology and an understanding of how coastal processes operate. This section will outline some key case studies of seawalls in chronological order and describe how they have performed in response to tsunami or ongoing natural processes and how effective they were in these situations. Analysing the successes and shortcomings of seawalls during severe natural events allows their weaknesses to be exposed, and areas become visible for future improvement.

Pondicherry

On December 26, 2004, towering waves of the 2004 Indian Ocean earthquake tsunami crashed against India's south-eastern coastline killing thousands. However, the former French colonial enclave of Pondicherry escaped unscathed. This was primarily due to French engineers who had constructed (and maintained) a massive stone seawall during the time when the city was a French colony. This 300-year-old seawall effectively kept Pondicherry's historic center dry even though tsunami waves drove water 24 ft (7.3 m) above the normal high-tide mark.

The barrier was initially completed in 1735 and over the years, the French continued to fortify the wall, piling huge boulders along its 1.25 mi (2 km) coastline to stop erosion from the waves pounding the harbor. At its highest, the barrier running along the water's edge reaches about 27 ft (8.2 m) above sea level. The boulders, some weighing up to a ton, are weathered black and brown. The sea wall is inspected every year and whenever gaps appear or the stones sink into the sand, the government adds more boulders to keep it strong (Allsop, 2002).

The Union Territory of Pondicherry recorded some 600 deaths from the huge tsunami waves that struck India's coast after the mammoth underwater earthquake (which measured 9.0 on the moment magnitude scale) off Indonesia, but most of those killed were fishermen who lived in villages beyond the artificial barrier which reinforces the effectiveness of seawalls.

Vancouver

The Vancouver Seawall is a stone seawall constructed around the perimeter of Stanley Park in Vancouver. The seawall was constructed initially as waves created by ships passing through the First Narrows were eroding the area between Prospect Point and Brockton Point. The Vancouver Seawall also exemplifies how seawalls can be utilised and valued for recreational activities and coastal sightseeing. A pedestrian, cycling and roller blading pathway exists on the seawall and has been extended far outside the parameters of Stanley Park. Construction of the seawall began in 1917, and since then this pathway has become one of the most used features of the park by both locals and tourists and now extends 22 km in total (Belyea & Ross, 1992). The construction of the seawall also provided employment for relief workers during the Great Depression and seamen from HMCS *Discovery* on Deadman's Island who were facing punishment detail in the 1950s (Steele, 1985).

Overall, the Vancouver Seawall is a prime example of how seawalls can simultaneously provide shoreline protection and a source of recreation which enhances human enjoyment of the coastal environment. It also illustrates that although shoreline erosion is a natural process, human activities, interactions with the coast and poorly planned shoreline development projects can accelerate natural erosion rates.

Japan

At least 43 percent of Japan's 29,751 km (18,486 mi) coastline is lined with concrete seawalls or other structures designed to protect the country against high waves, typhoons or even tsunamis. During the 2011 Tōhoku earthquake and tsunami, the seawalls in most areas were overwhelmed. In Kamaishi, 4-metre (13 ft) waves surmounted the seawall — the world's largest, erected a few years ago in the city's harbour at a depth of 63 m (207 ft), a length of 2 km (1.2 mi) and a cost of $1.5 billion — and eventually submerged the city center.

The risks of dependence on seawalls was most evident in the crisis at the Fukushima Dai-ichi and Fukushima Dai-ni nuclear power plants, both located along the coast close to the earthquake zone, as the tsunami washed over walls that were supposed to protect the plants. Arguably, the additional defense provided by the seawalls presented an extra margin of time for citizens to evacuate and also stopped some of the full force of energy which would have caused the wave to climb higher in the backs of coastal valleys. In contrast, the seawalls also acted in a negative way to trap water and delay its retreat.

The failure of the world's largest seawall, which cost $1.5 billion to construct, shows that building stronger sea walls to protect larger areas would have been even less cost-effective. In the case of the ongoing crisis at the nuclear power plants, higher and stronger sea walls should have been built if power plants were to be built at that site. Fundamentally, the devastation in coastal areas and a final death toll predicted to exceed 10,000 could push Japan to redesign its seawalls or consider more effective alternative methods of coastal protection for extreme events. Such hardened coastlines can also provide a false sense of security to property owners and local residents as evident in this situation.

Seawalls along the Japanese coast have also been criticized for cutting settlements off from the sea, make beaches unusable, presenting an eyesore, disturbing wildlife, and being unnecessary.

Rain Garden

Business parking lot that drains to a rain garden. The curb retains the asphalt pavement, yet lets water flow off the edges.

A rain garden is a planted depression or a hole that allows rainwater runoff from impervious urban areas, like roofs, driveways, walkways, parking lots, and compacted lawn areas, the opportunity to be absorbed. This reduces rain runoff by allowing stormwater to soak into the ground (as opposed to flowing into storm drains and surface waters which causes erosion, water pollution, flooding, and diminished groundwater). They should be designed for specific soils and climates. The purpose of a rain garden is to improve water quality in nearby bodies of water and to ensure that rainwater becomes available for plants as groundwater rather than being sent through stormwater drains straight out to sea. Rain gardens can cut down on the amount of pollution reaching creeks and streams by up to 30%.

Native and adapted plants are recommended for rain gardens because they are more tolerant of one's local climate, soil, and water conditions; have deep and variable root systems for enhanced water infiltration and drought tolerance; habitat value and diversity for local ecological communities; and overall sustainability once established. There can be trade-offs associated with using native plants, including lack of availability for some species, late spring emergence, short blooming season, and relatively slow establishment. The plants — a selection of wetland edge vegetation, such as wildflowers, sedges, rushes, ferns, shrubs and small trees — take up excess water flowing into the rain garden. Water filters through soil layers before entering the groundwater system. Root systems enhance infiltration, maintain or even augment soil permeability, provide moisture redistribution, and sustain diverse microbial populations involved in biofiltration. Also, through the process of transpiration, rain garden plants return water vapor to the atmosphere. A more wide-ranging definition covers all the possible elements that can be used to capture, channel, divert, and make the most of the natural rain and snow that falls on a property. The whole garden can become a rain garden, and each component of the whole can become a small-scale rain garden in itself.

Restoring the Water Cycle and Mitigating Urbanization

In developed areas, natural depressions where storm water would pool, are filled in. The surface of the ground is often leveled or paved. Storm water is directed into storm drains which often may cause overflows of combined sewer systems or poisoning, erosion or flooding of waterways receiving the storm water runoff. Redirected storm water is often warmer than the groundwater normally feeding a stream, and has been linked to upset in some aquatic ecosystems primarily through the reduction of dissolved oxygen (DO). Storm water runoff is also a source of a wide variety of pollutants washed off hard or compacted surfaces during rain events. These pollutants include volatile organic compounds, pesticides, herbicides, hydrocarbons and trace metals Rain gardens are designed to capture the initial flow of storm water and reduce the accumulation of toxins flowing directly into natural waterways through ground filtration. They also reduce energy consumption. For example, "the cumulative storage capacity of these rain gardens exceeds a conventional stormwater's system's by 10 times." The National Science Foundation, the United States Environmental Protection Agency, and a number of research institutions are presently studying the impact of augmenting rain gardens with materials capable of capture or chemical reduction of the pollutants to benign compounds.

Rain garden, SUNY College of Environmental Science and Forestry, Syracuse, New York

Rain gardens are often located near a building's roof drainpipe (with or without rainwater tanks). Most rain gardens are designed to be an endpoint of drainage with a capacity to percolate all incoming water through a series of soil or gravel layers beneath the surface plantings. A French drain may be used to direct a portion of the rainwater to an overflow location for heavier rain events. By reducing peak stormwater discharge, rain gardens extend hydraulic lag time and somewhat mimic the natural water cycle displaced by urban development and allow for groundwater recharge. While rain gardens always allow for restored groundwater recharge, and reduced stormwater volumes, they may also increase pollution unless remediation materials are included in the design of the filtration layers.

The primary challenge of rain garden design centers on calculating the types of pollutants and the acceptable loads of pollutants the rain garden's filtration system can handle during storm-water events. This challenge is specifically acute when a rain event occurs after a longer dry period. The initial storm water is often highly contaminated with the accumulated pollutants from dry periods. Rain garden designers have previously focused on finding robust native plants and encouraging adequate biofiltration, but recently have begun augmenting filtration layers with media specifically suited to chemically reduce redox of incoming pollutant streams.

Rain gardens are beneficial for many reasons: improve water quality by filtering runoff, provide localized flood control, are aesthetically pleasing, and provide interesting planting opportunities. They also encourage wildlife and biodiversity, tie together buildings and their surrounding environments in attractive and environmentally advantageous ways, and provide significant partial solutions to important environmental problems that affect us all.

A rain garden provides a way to use and optimize any rain that falls, reducing or avoiding the need for irrigation. They allow a household or building to deal with excessive rainwater runoff without burdening the public storm water systems. Rain gardens differ from retention basins, in that the water will infiltrate the ground within a day or two. This creates the advantage that the rain garden does not allow mosquitoes to breed.

History

The first rain gardens were created to mimic the natural water retention areas that occurred naturally before development of an area. The rain gardens for residential use were developed in 1990 in Prince George's County, Maryland, when Dick Brinker, a developer building a new housing subdivision had the idea to replace the traditional best management practices (BMP) pond with a bioretention area. He approached Larry Coffman, the county's Associate Director for Programs and Planning in the Department of Environmental Resources, with the idea. The result was the extensive use of rain gardens in Somerset, a residential subdivision which has a 300–400 sq ft (28–37 m^2) rain garden on each house's property. This system proved to be highly cost-effective. Instead of a system of curbs, sidewalks, and gutters, which would have cost nearly $400,000, the planted drainage swales cost $100,000 to install. This was also much more cost effective than building BMP ponds that could handle 2-, 10-, and 100-year storm events. Flow monitoring done in later years showed that the rain gardens have resulted in a 75–80% reduction in stormwater runoff during a regular rainfall event.

This is also referred to as Water Sensitive Urban Design (WSUD) in Australia, Sustainable Urban Drainage Systems or SUDS in the United Kingdom, and low impact development (LID) in the United States, and is cited by the EPA on their website as a success on the Stormwater Case Studies section of their website. This webpage has many links to information on Prince George's County's literature on implementing LID in a community.

Some *de facto* rain gardens predate their recognition by professionals as a significant LID tool. Any shallow garden depression implemented to capture and retain rain water within the garden so as to drain adjacent land without running off a property is at conception a rain garden — particularly if vegetation is maintained with recognition of its role in this function. Vegetated roadside swales, now promoted as "bioswales", remain the conventional drainage system in many parts of the world from long before extensive networks of concrete sewers became the conventional engineering practice in the industrialized world.

What is new about such technology is the emerging rigor of increasingly quantitative understanding of how such tools may make sustainable development possible. This is as true for wealthy developed communities retrofitting bioretention into built stormwater management systems, as for developing communities seeking a faster and more sustainable development path.

Characteristics

A home rain garden recently planted

A rain garden requires an area where water can collect and infiltrate, and plants to maintain infiltration rates, diverse microbe communities, and water holding capacity. Transpiration by growing plants accelerates soil drying between storms. This includes any plant extending roots to the garden area.

Simply adjusting the landscape so that downspouts and paved surfaces drain into existing gardens may be all that is needed because the soil has been well loosened and plants are well established. However, many plants do not tolerate saturated roots for long and often more water runs off one's roof than people realize. Often the required location and storage capacity of the garden area must be determined first. Rain garden plants are then selected to match the situation, not the other way around.

Soil and Drainage

When an area's soils are not permeable enough to allow water to drain and filter properly, the soil should be replaced and an underdrain installed. This bioretention mixture should typically contain 60% sand, 20% compost, and 20% topsoil, and there is a current trend to replace compost with biochar. Existing soil must be removed and replaced. Do not combine the sandy soil (bioretention) mixture with a surrounding soil that does not have high sand content. Otherwise, the clay particles will settle in between the sand particles and form a concrete-like substance, as demonstrated in a 1983 study. Deep plant roots also create additional channels for storm water to filter into the ground. Microbial populations feed off plant root secretions and break down carbon (such as in mulch or desiccated plant roots) to aggregate soil particles which increases infiltration rates. A five-year USGS study indicates that rain gardens in urban clay soils can be effective without the use of underdrains or replacement of native soils with the bioretention mix. Pre-installation infiltration rates should be at least .25 in/

hour, however. Type D soils will require an underdrain paired with the sandy soil mix in order to drain properly.

Sometimes a drywell with a series of gravel layers near the lowest spot in the rain garden will help facilitate percolation. However, a drywell placed at the lowest spot can become clogged with silt prematurely turning the garden into an infiltration basin defeating its purpose. Depression-focused recharge of polluted water into wells poses a serious threat and should be avoided. Similarly plans to install a rain garden near a septic system should be reviewed by a qualified engineer. The more polluted the water, the longer it must be retained in the soil for purification. This is often achieved by installing several smaller rain garden basins with soil deeper than the seasonal high water table. In some cases lined bioretention cells with subsurface drainage are used to retain smaller amounts of water and filter larger amounts without letting water percolate as quickly.

Rain gardens are at times confused with bioswales. Swales slope to a destination, while rain gardens do not; however, a bioswale may end with a rain garden. Drainage ditches may be handled like bioswales and even include rain gardens in series, saving time and money on maintenance. Part of a garden that nearly always has standing water is a water garden, wetland, or pond, and not a rain garden. Using the proper terminology ensures that the proper methods are used to achieve the desired results.

Plant Selection

Plants selected for use in a rain garden should tolerate both saturated and dry soil. Using native plants is generally encouraged. This way the rain garden may contribute to urban habitats for native butterflies, birds, and beneficial insects.

Well planned plantings require minimal maintenance to survive, and are compatible with adjacent land use. Trees under power lines, or that up-heave sidewalks when soils become moist, or whose roots seek out and clog drainage tiles can cause expensive damage.

Trees generally contribute most when located close enough to tap moisture in the rain garden depression, yet do not excessively shade the garden. That said, shading open surface waters can reduce excessive heating of habitat. Plants tolerate inundation by warm water for less time because heat drives out dissolved oxygen, thus a plant tolerant of early spring flooding may not survive summer inundation.

Rain Garden Projects

Australia

- Healthy Waterways Raingardens Program promotes a simple and effective form of stormwater treatment, and aims to raise peoples' awareness about how good stormwater management contributes to healthy waterways. The program encourages people to build rain gardens at home, and has achieved its target is to see 10,000 rain gardens built across Melbourne by 2013.

- Melbourne Water's database of Water Sensitive Urban Design projects, including 57 case studies relating to rain gardens/bioretention systems. Melbourne Water is the Victorian State Government agency responsible for managing Melbourne's water supply catchments.

- Water By Design is a capacity building program that supports the uptake of Water Sensitive Urban Design, including rain gardens, in South East Queensland. It was established by the South East Queensland Healthy Waterways Partnership in 2005, as an integral component of the SEQ Healthy Waterways Strategy.

United Kingdom

- The Wildfowl and Wetlands Trust's London Wetland Centre includes a rain garden designed by Nigel Dunnett.

- Islington London Borough Council commissioned sustainable drainage consultants Robert Bray Associates to design a pilot rain garden in the Ashby Grove development which was completed in 2011. This raingarden is fed from a typical modest domestic roof catchment area of $30m^2$ and is designed to demonstrate how simple and cost effective domestic rain gardens are to install. Monitoring apparatus was built into the design to allow Middlesex University to monitor water volumes, water quality and soil moisture content. The rain garden basin is 300mm deep and has a storage capacity of $2.17m^3$ which is just over the volume required to store runoff from the roof catchment in a 1 in 100 storm plus 30% allowance for climate change.

- The Day Brook Rain Garden Project has introduced a number of rain gardens into an existing residential street in Sherwood, Nottingham

United States of America

- The 12,000 rain garden campaign for Puget Sound is coordinating efforts to build 12,000 rain gardens in the Puget Sound Basin of Western Washington by 2016. The 12,000 rain gardens website provides information and resources for the general public, landscape professionals, municipal staff, and decision makers. By providing access to the best current guidance, easy-to-use materials, and a network of trained "Rain Garden Mentor" Master Gardeners, this campaign seeks to capture and cleanse over 200 Million gallons of polluted runoff each year, and thereby significantly improve Puget Sound's water quality.

- Maplewood, Minnesota has implemented a policy of encouraging residents to install rain gardens. Many neighborhoods had swales added to each property, but installation of a garden at the swale was voluntary. The project was a partnership between the City of Maplewood, University of Minnesota Department of Landscape Architecture, and the Ramsey Washington Metro Watershed District. A focus group was held with residents and published so that other communities could use it as a resource when planning their own rain garden projects.

- In Seattle, a prototype project, used to develop a plan for the entire city, was constructed in 2003. Called *SEA Street,* for Street Edge Alternatives, it was a drastic facelift of a residential street. The street was changed from a typical linear path to a gentle curve, narrowed, with large rain gardens placed along most of the length of the street. The street has 11% less impervious surface than a regular street. There are 100 evergreen trees and 1100 shrubs along this 3-block stretch of road, and a 2-year study found that the amount of stormwater which leaves the street has been reduced by 99%.
- 10,000 Rain Gardens is a public initiative in the Kansas City, Missouri metro area. Property owners are encouraged to create rain gardens, with an eventual goal of 10,000 individual gardens.
- The West Michigan Environmental Action Council has established Rain Gardens of West Michigan as an outreach water quality program. Also in Michigan, the Southeastern Oakland County Water Authority has published a pamphlet to encourage residents to add a rain garden to their landscapes in order to improve the water quality in the Rouge River watershed. In Washtenaw County, homeowners can volunteer for the Water Resources Commissioner's Rain Garden program, in which volunteers are annually selected for free professional landscape design. The homeowners build the gardens themselves as well as pay for landscaping material. Photos of the gardens as well as design documents and drainage calculations are available online.
- The city of Portland, Oregon, has established a Clean River Rewards program, to encourage residents to disconnect downspouts from the city's combined sewer system and create rain gardens. Workshops, discounts on storm water bills, and web resources are offered.
- In Delaware, several rain gardens have been created through the work of the University of Delaware Water Resources Agency, and environmental organizations, such as the Appoquinimink River Association.
- In New Jersey, the Rutgers Cooperative Extension Water Resources Program has already installed over 125 demonstration rain gardens in suburban and urban areas. The Water Resources Program has begun to focus on using rain gardens as green infrastructure in urban areas, such as Camden and Newark to help prevent localized flooding, combined sewer overflows, and to improve water quality. The Water Resources Program has also revised and produced

a rain garden manual in collaboration with The Native Plant Society of New Jersey.

Land Reclamation

Reclaiming in Perth, Australia 1964

Land reclamation, usually known as reclamation, and also known as land fill is the process of creating new land from ocean, riverbeds, or lake beds. The land reclaimed is known as reclamation ground or land fill.

In a number of other jurisdictions, including parts of the United States, the term “reclamation” can refer to returning disturbed lands to an improved state. In Alberta, Canada, for example, reclamation is defined by the provincial government as “The process of reconverting disturbed land to its former or other productive uses.” In Oceania it is frequently referred to as land rehabilitation.

Methods

Land reclamation can be achieved with a number of different methods. The most simple method involves simply filling the area with large amounts of heavy rock and/or cement, then filling with clay and dirt until the desired height is reached. The process is called “infilling” and the material used to fill the space is generically called “infill”. Draining of submerged wetlands is often used to reclaim land for agricultural use. Deep cement mixing is used typically in situation in which the material displaced by either dredging or draining may be contaminated and hence needs to be contained.

Habitation

The Flevopolder in the Netherlands, reclaimed from the IJsselmeer, is the largest reclaimed artificial island in the world.

Land Reclamation in the Beirut Central District

The whole district of Fontvieille, Monaco was reclaimed from the sea

The creation of new land was for the need of human activities.

Notable examples include:

Asia

- Much of the coastlines of Mumbai, India. It took over 150 years to join the original seven islands of Mumbai. These seven islands were lush green thickly wooded, and dotted with 22 hills, with the Arabian Sea washing through them at high tide.The original island of Mumbai was only 24 km long and 4 km wide from Dongri to Malabar Hill (at its broadest point) and the other six were Colaba , Old Woman's island , Mahim , Parel , Worli , Mazgaon .
- Much of the coastlines of Mainland China, Hong Kong, North Korea and South Korea. It is estimated that nearly 65% of tidal flats around the Yellow Sea have been reclaimed.
- Inland lowlands in the Yangzi valley, China, including the areas of important cities like Shanghai and Wuhan.
- Much of the coastline of Karachi, Pakistan.
- A part of the Hamad International Airport in Qatar, around 36 square kilometres (14 sq mi).
- The entire island of The Pearl-Qatar situated in West Bay (Doha), Qatar.
- Haikou Bay, Hainan Province, China, where the west side of Haidian Island is being extended, and off the coast of Haikou City, where new land for a marina is being created
- The Cotai Strip in Macau, where most of the major casinos are located
- Nagoya Centrair Airport, Japan
- Incheon International Airport, Korea
- Beirut Central District, Lebanon
- The southern Chinese city of Shenzhen
- The shore of Manila Bay in the Philippines, especially along Metro Manila, has attracted major developments such as the Mall of Asia Complex, Entertainment City and the Cultural Centre of the Philippines Complex.
- The city-state of Singapore, where land is in short supply, is also famous for its efforts on land reclamation.
- The Palm Islands, The World and hotel Burj al-Arab off Dubai in the United Arab Emirates
- The Yas Island in Abu Dhabi, UAE.
- Hulhumalé Island, Maldives. It is one of the six divisions of Malé City.

Europe

- Large parts of the Netherlands
- Parts of Dublin, Ireland
- Most of Belfast Harbour and areas of Belfast, Northern Ireland
- Parts of Saint Petersburg, Russia, such as the Marine Facade
- Helsinki (of which the major part of the city center is built on reclaimed land)
- Barceloneta area, Barcelona, in Spain
- The port of Zeebrugge in Belgium
- The southwestern residential area in Brest, Belarus
- Most of Fontvieille, Monaco
- Parts surrounding Port Hercules in La Condamine, Monaco
- The airport peninsula, the industrial area of Cornigliano, the PSA container terminal and other parts of the port in Genoa, Italy
- The Fens in East Anglia
- Venice, Italy
- Rione Orsini, part of Borgo Santa Lucia, Naples

Africa

- The Foreshore in Cape Town
- The Hassan II Mosque in Morocco is built on reclaimed land.
- The Eko Atlantic in Lagos, Nigeria.

Americas

- Large parts of Rio de Janeiro, most notably several blocks in the new docks area, the entire Flamengo Park and the neighborhood of Urca
- Parts of New Orleans (which is partially built on land that was once swamp)
- Parts of Montevideo, Uruguay, Rambla Sur and several projects still going on in Montevideo's Bay.
- Much of the urbanized area adjacent to San Francisco Bay, including most of San Francisco's waterfront and Financial District, the Port of Oakland, and large portions of the city of Alameda has been reclaimed from the bay.

- Mexico City (which is situated at the former site of Lake Texcoco); the chinampas are a famous example.
- Parts of Panama City urban and street development are based on reclaimed land, using material extracted from Panama Canal excavations.
- The Chicago shoreline
- The Northwestern University Lakefill, part of the campus of Northwestern University in Evanston, Illinois
- Back Bay, Boston, Massachusetts
- Battery Park City, Manhattan
- Liberty State Park, Jersey City
- The Toronto Islands, Leslie Street Spit, and the waterfront in Toronto
- Part of Nuns' Island and all of Île Notre-Dame in Montreal
- The Cinta Costera, in Panama City, Panama
- The Potter's Cay in Nassau, The Bahamas was connected to the island of New Providence
- The shore of Nassau, The Bahamas especially along East Bay street.

One of the earliest large scale projects was the Beemster Polder in the Netherlands, realized in 1612 adding 70 square kilometres (27 sq mi) of land. In Hong Kong the Praya Reclamation Scheme added 20 to 24 hectares (50 to 60 acres) of land in 1890 during the second phase of construction. It was one of the most ambitious projects ever taken during the Colonial Hong Kong era. Some 20% of land in the Tokyo Bay area has been reclaimed, most notably Odaiba artificial island. Le Portier, Monaco and Gibraltar are also expanding due to land reclamation. The city of Rio de Janeiro was largely built on reclaimed land, as was Wellington, New Zealand.

Artificial islands are an example of land reclamation. Creating an artificial island is an expensive and risky undertaking. It is often considered in places with high population density and a scarcity of flat land. Kansai International Airport (in Osaka) and Hong Kong International Airport are examples where this process was deemed necessary. The Palm Islands, The World and hotel Burj al-Arab off Dubai in the United Arab Emirates are other examples of artificial islands (although there is yet no real "scarcity of land" in Dubai), as well as the Flevopolder in the Netherlands which is the largest artificial island in the world.

Agriculture

Agriculture was a drive for land reclamation before industrialisation. In South China, farmers reclaimed paddy fields by enclosing an area with a stone wall on the sea shore near river mouth or river delta. The species of rice that grow on these grounds are more salt tolerant. Another use of such enclosed land is creation of fish ponds. It is commonly seen on the Pearl River Delta and Hong Kong. These reclamation also attracts species of migrating birds.

Land reclamation in progress in Bingzhou (丙州) Peninsula (formerly, island) of the Dongzui Bay (东咀港). Tong'an District, Xiamen, China

A related practice is the draining of swampy or seasonally submerged wetlands to convert them to farmland. While this does not create new land exactly, it allows commercially productive use of land that would otherwise be restricted to wildlife habitat. It is also an important method of mosquito control.

Even in the post-industrial age, there have been land reclamation projects intended for increasing available agricultural land. For example, the village of Ogata in Akita, Japan, was established on land reclaimed from Lake Hachirogata (Japan's second largest lake at the time) starting in 1957. By 1977, the amount of land reclaimed totalled 172.03 square kilometres (66.42 sq mi).

Beach Restoration

Beach rebuilding is the process of repairing beaches using materials such as sand or mud from inland. This can be used to build up beaches suffering from beach starvation or erosion from longshore drift. It stops the movement of the original beach material through longshore drift and retains a natural look to the beach. Although it is not a long-lasting solution, it is cheap compared to other types of coastal defences.

Landfill

As human overcrowding of developed areas intensified during the 20th century, it has become important to develop land re-use strategies for completed landfills. Some of the most common usages are for parks, golf courses and other sports fields. Increasingly, however, office buildings and industrial uses are made on a completed landfill. In these latter uses, methane capture is customarily carried out to minimize explosive hazard within the building.

An example of a Class A office building constructed over a landfill is the Dakin Building at Sierra Point, Brisbane, California. The underlying fill was deposited from 1965 to 1985, mostly consisting of construction debris from San Francisco and some municipal wastes. Aerial photographs prior to 1965 show this area to be tidelands of the San Francisco Bay. A clay cap was constructed over the debris prior to building approval.

A notable example is Sydney Olympic Park, the primary venue for the 2000 Summer Olympic Games, which was built atop an industrial wasteland that included landfills.

Another strategy for landfill is the incineration of landfill trash at high temperature via the plasma-arc gasification process, which is currently used at two facilities in Japan, and will be used at a planned facility in St. Lucie County, Florida.

Environmental Impact

Draining wetlands for ploughing, for example, is a form of habitat destruction. In some parts of the world, new reclamation projects are restricted or no longer allowed, due to environmental protection laws.

Parts (highlighted in brown) of the San Francisco Bay were reclaimed from wetlands for urban use.

Environmental Legislation

The State of California created a state commission, the San Francisco Bay Conservation and Development Commission, in 1965 to protect San Francisco Bay and regulate development near its shores. The commission was created in response to growing concern over the shrinking size of the bay.

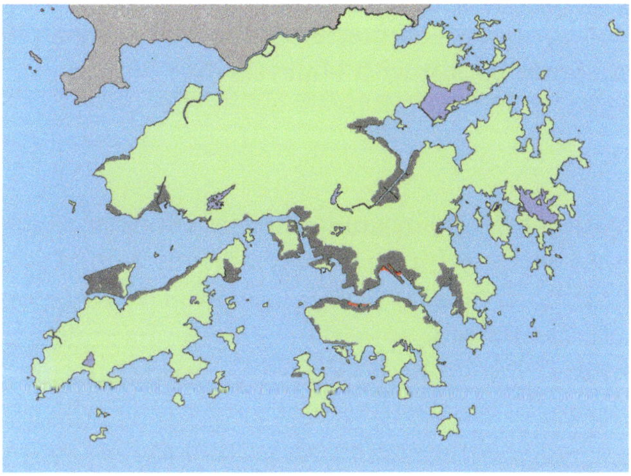

A map of reclaimed land in Hong Kong: Grey (built), red (proposed or under development). Many of the urban areas of Hong Kong are on reclaimed land.

Hong Kong legislators passed the Protection of the Harbour Ordinance, proposed by the Society for Protection of the Harbour, in 1997 in an effort to safeguard the increasingly threatened Victoria Harbour against encroaching land development. Several large reclamation schemes at Green Island, West Kowloon, and Kowloon Bay were subsequently shelved, and others reduced in size.

Dangers

Reclaimed land is highly susceptible to soil liquefaction during earthquakes, which can amplify the amount of damage that occurs to buildings and infrastructure. Subsidence is another issue, both from soil compaction on filled land, and also when wetlands are enclosed by levees and drained to create Polders. Drained marshes will eventually sink below the surrounding water level, increasing the danger from flooding.

Land Amounts Added

- Bangladesh – about 110 square kilometres (42 sq mi) in total and has 12,000 square kilometres (4,600 sq mi) potential (8% of total area) up to 12 metres (39 ft) depth in the territorial sea area.
- Netherlands – about 1/6 of the entire country, or about 7,000 square kilometres (2,700 sq mi) in total, has been reclaimed from the sea, lakes, marshes and swamps.

- South Korea – As of 2006, 38 percent or 1,550 square kilometres (600 sq mi) of coastal wetlands reclaimed, including 400 square kilometres (150 sq mi) at Saemangeum. Songdo International Business district, the largest private development in history, is a large-scale reclamation project built entirely on tidal mudflats.
- Singapore – 20% of the original size or 135 square kilometres (52 sq mi). As of 2003, plans for 99 square kilometres (38 sq mi) more are to go ahead, despite the fact that disputes persist with Malaysia over Singapore's extensive land reclamation works.
- Hong Kong

  Praya Reclamation Scheme began in the late 1860s and consisted of two stages totaling 20 to 24 hectares (50 to 60 acres). Hong Kong Disneyland, Hong Kong International Airport, and its predecessor, Kai Tak Airport, were all built on reclaimed land. In addition, much reclamation has taken place in prime locations on the waterfront on both sides of Victoria Harbour. This has raised environmental issues of the protection of the harbour which was once the source of prosperity of Hong Kong, traffic congestion in the Central district, as well as the collusion of the Hong Kong Government with the real estate developers in the territory.

  In addition, as the city expands, new towns in different decades were mostly built on reclaimed land, such as Tuen Mun, Tai Po, Shatin-Ma On Shan, West Kowloon, Kwun Tong and Tseung Kwan O.
- Macau – 170% of the original size or 17 square kilometres (6.6 sq mi)
- Mumbai – An archipelago of originally seven separate islands were joined together by land reclamation over a span of five centuries. This was done to develop Mumbai as a harbour city.
- Monaco – 0.41 square kilometres (0.16 sq mi) out of 2.05 square kilometres (0.79 sq mi), or 1/5 of Monaco comes from land taken from the sea, mainly in the neighborhoods of Fontvieille, La Condamine, and Larvotto/Bas Moulins.
- Tokyo Bay, Japan – 249 square kilometres (96 sq mi) including the entirety of Odaiba artificial island.
- Kobe, Japan – 23 square kilometres (8.9 sq mi) (1995).
- Bahrain – 76.3% of original size of 410 square kilometres (160 sq mi) (1931–2007).
- New Zealand – significant areas of land totalling several hundred hectares have been reclaimed along the harbourfront of Auckland, Wellington and Dunedin.

In Dunedin – which in its early days was nicknamed "Mudedin" – around 2.5 square kilometres (0.97 sq mi), including much of the inner city and suburbs of Dunedin North, South Dunedin and Andersons Bay is reclaimed from the Otago Harbour, and a similar area in the suburbs of St Clair and St Kilda is reclaimed swampland.

- Dubai, UAE - Dubai has a total of four reclaimed islands (the palm Jumeirah, Jebal Ali, The Burj al Arab island and The World Islands), with a fifth under construction (the Palm Deira)

Hydraulic Engineering

Hydraulic engineering as a sub-discipline of civil engineering is concerned with the flow and conveyance of fluids, principally water and sewage. One feature of these systems is the extensive use of gravity as the motive force to cause the movement of the fluids. This area of civil engineering is intimately related to the design of bridges, dams, channels, canals, and levees, and to both sanitary and environmental engineering.

Hydraulic Flood Retention Basin (HFRB)

View from Church Span Bridge, Bern, Switzerland

Riprap lining a lake shore

Hydraulic engineering is the application of fluid mechanics principles to problems dealing with the collection, storage, control, transport, regulation, measurement, and use of water. Before beginning a hydraulic engineering project, one must figure out how much water is involved. The hydraulic engineer is concerned with the transport of sediment by the river, the interaction of the water with its alluvial boundary, and the occurrence of scour and deposition. "The hydraulic engineer actually develops conceptual designs for the various features which interact with water such as spillways and outlet works for dams, culverts for highways, canals and related structures for irrigation projects, and cooling-water facilities for thermal power plants."

Fundamental Principles

A few examples of the fundamental principles of hydraulic engineering include fluid mechanics, fluid flow, behavior of real fluids, hydrology, pipelines, open channel hydraulics, mechanics of sediment transport, physical modeling, hydraulic machines, and drainage hydraulics.

Fluid Mechanics

Fundamentals of Hydraulic Engineering defines hydrostatics as the study of fluids at rest. In a fluid at rest, there exists a force, known as pressure, that acts upon the fluid's surroundings. This pressure, measured in N/m^2, is not constant throughout the body of fluid. Pressure, p, in a given body of fluid, increases with an increase in depth. Where the upward force on a body acts on the base and can be found by equation:

$$p = \rho g y$$

where,

ρ = density of water

g = specific gravity

y = depth of the body of liquid

Rearranging this equation gives you the pressure head $p/\rho g = y$. Four basic devices for pressure measurement are a piezometer, manometer, differential manometer, Bourdon gauge, as well as an inclined manometer.

As Prasuhn states:

> On undisturbed submerged bodies, pressure acts along all surfaces of a body in a liquid, causing equal perpendicular forces in the body to act against the pressure of the liquid. This reaction is known as equilibrium. More advanced applications of pressure are that on plane surfaces, curved surfaces, dams, and quadrant gates, just to name a few.

Behavior of Real Fluids

Real and Ideal Fluids

The main difference between an ideal fluid and a real fluid is that for ideal flow $p_1 = p_2$ and for real flow $p_1 > p_2$. Ideal fluid is incompressible and has no viscosity. Real fluid has viscosity. Ideal fluid is only an imaginary fluid as all fluids that exist have some viscosity.

Viscous Flow

A viscous fluid will deform continuously under a shear force, whereas an ideal fluid doesn't deform.

Laminar Flow and Turbulence

The various effects of disturbance on a viscous flow are stable, transition and unstable.

Bernoulli's Equation

For an ideal fluid, Bernoulli's equation holds along streamlines.

$$p/\rho g + u^2/2g = p_1/\rho g + u_1^2/2g = p_2/\rho g + u_2^2/2g$$

Boundary Layer

Assuming a flow is bounded on one side only, and that a rectilinear flow passing over a stationary flat plate which lies parallel to the flow, the flow just upstream of the plate has a uniform velocity. As the flow comes into contact with the plate, the layer of fluid actually 'adheres' to a solid surface. There is then a considerable shearing action between the layer of fluid on the plate surface and the second layer of fluid. The second layer is therefore forced to decelerate (though it is not quite brought to rest), creating a shearing action with the third layer of fluid, and so on. As the fluid passes further along the plate, the zone in which shearing action occurs tends to spread further outwards.

This zone is known as the 'boundary layer'. The flow outside the boundary layer is free of shear and viscous-related forces so it is assumed to act like an ideal fluid. The intermolecular cohesive forces in a fluid are not great enough to hold fluid together. Hence a fluid will flow under the action of the slightest stress and flow will continue as long as the stress is present. The flow inside the layer can be either viscous or turbulent, depending on Reynolds number.

Applications

Common topics of design for hydraulic engineers include hydraulic structures such as dams, levees, water distribution networks, water collection networks, sewage collection networks, storm water management, sediment transport, and various other topics related to transportation engineering and geotechnical engineering. Equations developed from the principles of fluid dynamics and fluid mechanics are widely utilized by other engineering disciplines such as mechanical, aeronautical and even traffic engineers.

Related branches include hydrology and rheology while related applications include hydraulic modeling, flood mapping, catchment flood management plans, shoreline management plans, estuarine strategies, coastal protection, and flood alleviation.

History

Antiquity

Earliest uses of hydraulic engineering were to irrigate crops and dates back to the Middle East and Africa. Controlling the movement and supply of water for growing food has been used for many thousands of years. One of the earliest hydraulic machines, the water clock was used in the early 2nd millennium BC. Other early examples of using gravity to move water include the Qanat system in ancient Persia and the very similar Turpan water system in ancient China as well as irrigation canals in Peru.

In ancient China, hydraulic engineering was highly developed, and engineers constructed massive canals with levees and dams to channel the flow of water for irrigation, as well as locks to allow ships to pass through. Sunshu Ao is considered the first Chinese hydraulic engineer. Another important Hydraulic Engineer in China, Ximen Bao was credited of starting the practice of large scale canal irrigation during the Warring States period (481 BC-221 BC), even today hydraulic engineers remain a respectable position in China. Before becoming General Secretary of the Communist Party of China in 2002, Hu Jintao was a hydraulic engineer and holds an engineering degree from Tsinghua University

Eupalinos of Megara, was an ancient Greek engineer who built the Tunnel of Eupalinos on Samos in the 6th century BC, an important feat of both civil and hydraulic engineering. The civil engineering aspect of this tunnel was the fact that it was dug from both ends which required the diggers to maintain an accurate path so that the two tunnels met and that the entire effort maintained a sufficient slope to allow the water to flow.

Hydraulic engineering was highly developed in Europe under the aegis of the Roman Empire where it was especially applied to the construction and maintenance of aqueducts to supply water to and remove sewage from their cities. In addition to supplying the needs of their citizens they used hydraulic mining methods to prospect and extract alluvial gold deposits in a technique known as hushing, and applied the methods to other ores such as those of tin and lead.

In the 15th century, the Somali Ajuran Empire was the only hydraulic empire in Africa. As a hydraulic empire, the Ajuran State monopolized the water resources of the Jubba and Shebelle Rivers. Through hydraulic engineering, it also constructed many of the limestone wells and cisterns of the state that are still operative and in use today. The rulers developed new systems for agriculture and taxation, which continued to be used in parts of the Horn of Africa as late as the 19th century.

Further advances in hydraulic engineering occurred in the Muslim world between the 8th to 16th centuries, during what is known as the Islamic Golden Age. Of particular importance was the 'water management technological complex' which was central to the Islamic Green Revolution and, by extension, a precondition for the emergence of modern technology. The various components of this 'toolkit' were developed in different parts of the Afro-Eurasian landmass, both within and beyond the Islamic world. However, it was in the medieval Islamic lands where the technological complex was assembled and standardized, and subsequently diffused to the rest of the Old World. Under the rule of a single Islamic Caliphate, different regional hydraulic technologies were assembled into "an identifiable water management technological complex that was to have a global impact." The various components of this complex included canals, dams, the qanat system from Persia, regional water-lifting devices such as the noria, shaduf and screwpump from Egypt, and the windmill from Islamic Afghanistan. Other original Islamic developments included the saqiya with a flywheel effect from Islamic Spain, the reciprocating suction pump and crankshaft-connecting rod mechanism from Iraq, the geared and hydropowered water supply system from Syria, and the water purification methods of Islamic chemists.

Modern Times

In many respects the fundamentals of hydraulic engineering haven't changed since ancient times. Liquids are still moved for the most part by gravity through systems of canals and aqueducts, though the supply reservoirs may now be filled using pumps. The need for water has steadily increased from ancient times and the role of the hydraulic engineer is a critical one in supplying it. For example, without the efforts of people like William Mulholland the Los Angeles area would not have been able to grow as it has because it simply doesn't have enough local water to support its population. The same is true for many of our world's largest cities. In much the same way, the central valley of California could not have become such an important agricultural region without effective water management and distribution for irrigation. In a somewhat parallel way to what happened in California the creation of the Tennessee Valley Authority(TVA) brought work and prosperity to the South by building dams to generate cheap electricity and control flooding in the region, making rivers navigable and generally modernizing life in the region.

Leonardo da Vinci (1452–1519) performed experiments, investigated and speculated on waves and jets, eddies and streamlining. Isaac Newton (1642–1727) by formulating the laws of motion and his law of viscosity, in addition to developing the calculus, paved the way for many great developments in fluid mechanics. Using Newton's laws of motion, numerous 18th-century mathematicians solved many frictionless (zero-viscosity) flow problems. However, most flows are dominated by viscous effects, so engineers of the 17th and 18th centuries found the inviscid flow solutions unsuitable, and by experimentation they developed empirical equations, thus establishing the science of hydraulics.

Late in the 19th century, the importance of dimensionless numbers and their relationship to turbulence was recognized, and dimensional analysis was born. In 1904 Ludwig Prandtl published a key paper, proposing that the flow fields of low-viscosity fluids be divided into two zones, namely a thin, viscosity-dominated boundary layer near solid surfaces, and an effectively inviscid outer zone away from the boundaries. This concept explained many former paradoxes, and enabled subsequent engineers to analyze far more complex flows. However, we still have no complete theory for the nature of turbulence, and so modern fluid mechanics continues to be combination of experimental results and theory.

The modern hydraulic engineer uses the same kinds of computer-aided design (CAD) tools as many of the other engineering disciplines while also making use of technologies like computational fluid dynamics to perform the calculations to accurately predict flow characteristics, GPS mapping to assist in locating the best paths for installing a system and laser-based surveying tools to aid in the actual construction of a system.

Environmental Engineering

Environmental engineering is the branch of engineering concerned with the application of scientific and engineering principles for protection of human populations from the effects of adverse environmental factors; protection of environments, both local and global, from the potentially deleterious effects of natural and human activities; and improvement of environmental quality.

Environmental engineering can also be described as a branch of applied science and technology that addresses the issues of energy preservation, production asset and control of waste from human and animal activities. Furthermore, it is concerned with finding plausible solutions in the field of public health, such as waterborne diseases, implementing laws which promote adequate sanitation in urban, rural and recreational areas. It involves waste water management, air pollution control, recycling, waste disposal, radiation protection, industrial hygiene, environmental sustainability, public health and environmental engineering law. It also includes studies on the environmental impact of proposed construction projects.

Environmental engineers study the effect of technological advances on the environment. To do so, they conduct studies on hazardous-waste management to evaluate the significance of such hazards, advise on treatment and containment, and develop regulations to prevent mishaps. Environmental engineers design municipal water supply and industrial wastewater treatment systems. They address local and worldwide environmental issues such as the effects of acid rain, global warming, ozone depletion, water pollution and air pollution from automobile exhausts and industrial sources.

At many universities environmental engineering programs are offered at either the department of civil engineering or the department of chemical engineering at engineering faculties. Environmental "civil" engineers focus on hydrology, water resources management, bioremediation, and water treatment plant design. Environmental "chemical" engineers, on the other hand, focus on environmental chemistry, advanced air and water treatment technologies and separation processes.

Additionally, engineers are more frequently obtaining specialized training in law (J.D.) and are utilizing their technical expertise in the practices of environmental engineering law.

Most jurisdictions also impose licensing and registration requirements.

Development

Ever since people first recognized that their health is related to the quality of their environment, they have applied principles to attempt to improve the quality of their environment. The ancient Indian Harappan civilization utilized early sewers in some cities more than 5000 years ago. The Romans constructed aqueducts to prevent drought and to create a clean, healthful water supply for the metropolis of Rome. In the 15th century, Bavaria created laws restricting the development and degradation of alpine country that constituted the region's water supply.

The field emerged as a separate environmental discipline during the middle third of the 20th century in response to widespread public concern about water and pollution and increasingly extensive environmental quality degradation. However, its roots extend back to early efforts in public health engineering. Modern environmental engineering began in London in the mid-19th century when Joseph Bazalgette designed the first major sewerage system that reduced the incidence of waterborne diseases such as cholera. The introduction of drinking water treatment and sewage treatment in industrialized countries reduced waterborne diseases from leading causes of death to rarities.

In many cases, as societies grew, actions that were intended to achieve benefits for those societies had longer-term impacts which reduced other environmental qualities. One example is the widespread application of the pesticide DDT to control agricultural pests in the years following World War II. While the agricultural benefits were outstanding and crop yields increased dramatically thus reducing world hunger substantially, and malaria was controlled better than it ever had been, numerous species were brought to the verge of extinction due to the impact of the DDT on their reproductive cycles. The story of DDT as vividly told in Rachel Carson's Silent Spring (1962) is considered to be the birth of the modern environmental movement and of the modern field of "environmental engineering."

Conservation movements and laws restricting public actions that would harm the environment have been developed by various societies for millennia. Notable examples are the laws decreeing the construction of sewers in London and Paris in the 19th century and the creation of the U.S. national park system in the early 20th century.

Scope

Solid Waste Management

Environmental Impact Assessment and Mitigation

Scientists have air pollution dispersion models to evaluate the concentration of a pollutant at a receptor or the impact on overall air quality from vehicle exhausts and industrial flue gas stack emissions. To some extent, this field overlaps the desire to decrease carbon dioxide and other greenhouse gas emissions from combustion processes. They apply scientific and engineering principles to evaluate if there are likely to be any adverse impacts to water quality, air quality, habitat quality, flora and fauna, agricultural capacity, traffic impacts, social impacts, ecological impacts, noise impacts, visual (landscape) impacts, etc. If impacts are expected, they then develop mitigation measures to limit or prevent such impacts. An example of a mitigation measure would be the creation of wetlands in a nearby location to mitigate the filling in of wetlands necessary for a road development if it is not possible to reroute the road.

In the United States, the practice of environmental assessment was formally initiated on January 1, 1970, the effective date of the National Environmental Policy Act (NEPA). Since that time, more than 100 developing and developed nations either have planned specific analogous laws or have adopted procedure used elsewhere. NEPA is applicable to all federal agencies in the United States.

Water Supply and Treatment

Sewage treatment plant, Australia

They evaluate the water balance within a watershed and determine the available water supply, the water needed for various needs in that watershed, the seasonal cycles of water movement through the watershed and they develop systems to store, treat, and convey water for various uses. Water is treated to achieve water quality objectives for the end uses. In the case of a potable water supply, water is treated to minimize the risk of infectious disease transmission, the risk of non-infectious illness, and to create a palatable water flavor. Water distribution systems are designed and built to provide adequate water pressure and flow rates to meet various end-user needs such as domestic use, fire suppression, and irrigation.

Wastewater Treatment

There are numerous wastewater treatment technologies. A wastewater treatment train can consist of a primary clarifier system to remove solid and floating materials, a secondary treatment system consisting of an aeration basin followed by flocculation and sedimentation or an activated sludge system and a secondary clarifier, a tertiary biological nitrogen removal system, and a final disinfection process. The aeration basin/activated sludge system removes organic material by growing bacteria (activated sludge). The secondary clarifier removes the activated sludge from the water. The tertiary system, although not always included due to costs, is becoming more prevalent to remove nitrogen and phosphorus and to disinfect the water before discharge to a surface water stream or ocean outfall.

Water pollution

Air Pollution Management

Scientists have developed air pollution dispersion models to evaluate the concentration of a pollutant at a receptor or the impact on overall air quality from vehicle exhausts and industrial flue gas stack emissions. To some extent, this field overlaps the desire to decrease carbon dioxide and other greenhouse gas emissions from combustion processes.

Environmental Protection Agency

The U.S. Environmental Protection Agency (EPA) is one of the many agencies that work with environmental engineers to solve key issues. An important component of EPA's mission is to protect and improve air, water, and overall environmental quality in order to avoid or mitigate the consequences of harmful effects.

Ecological engineering for sustainable agriculture in arid and semiarid West African regions.

Ecological engineering offers new alternatives for the management of agricultural systems that are more tailored to the ever-changing social and environmental necessities in these regions. This requires managing the complexity of agrosystems, while striving to mimic the functioning of natural ecosystems of West African drylands and taking advantage of traditional practices and local know-how resulting from a long process of adaptation to environmental constraints.

1. Acting on biodiversity. Biodiversity is essential to the productivity of ecosystems and their temporal stability under the impact of external disturbances. Several ecological processes related to biodiversity may be intensified for the benefit of agrosilvopastoral systems: promoting diversity and soil microorganism activity to benefit plants, associating and utilizing the mutual benefits of plants

2. Utilizing organic matter and nutrient cycles. The productivity of agrosystems with low chemical input use in dryland regions is primarily based on efficient organic resource management, and in turn on the nutrient and energy flows they induce. It is thus possible to intervene at several levels: enhancing crop-livestock farming integration to preserve natural resources, restoring the biological activity of soils via specific organic inputs, supplying nutrients to plants locally.

3. Enhancing available water use. Water supplies are limited and irregular in dryland areas. Current management of these supplies—which involves capturing rainwater and surface runoff—could be improved in several ways: adapting to erratic rainfall or drought risks by focusing on: (i) the organization of the farm and community (farm plot patterns in association with the random rainfall distribution, etc.), and on (ii) cropping techniques to reduce crop water needs (plant choices, weeding, etc.), preserving water in crop fields by hampering runoff, accounting for the essential role of trees regarding soil and water in drylands.

4. Managing landscapes and associated ecological processes. Ecological crop pest regulation by their natural enemies is one ecosystem service provided by biodiversity. Better pest management could be considered in association with promoting biodiversity at different scales, e.g. from the plant to the landscape.

Education

Courses aimed at developing graduates with specific skills in environmental systems or environmental technology are becoming more common and fall into broad classes:

- Mechanical engineering courses oriented towards designing machines and mechanical systems for environmental use such as water treatment facilities, pumping stations, garbage segregation plants and other mechanical facilities;
- Environmental engineering or environmental systems courses oriented towards a civil engineering approach in which structures and the landscape are constructed to blend with or protect the environment;
- Environmental chemistry, sustainable chemistry or environmental chemical engineering courses oriented towards understanding the effects (good and bad) of chemicals in the environment. Focus on mining processes, pollutants and commonly also cover biochemical processes;
- Environmental technology courses oriented towards producing electronic or electrical graduates capable of developing devices and artifacts able to monitor, measure, model and control environmental impact, including monitoring and managing energy generation from renewable sources.

Prominent Environmental Engineers

- Robert A. Gearheart
- Paul V. Roberts
- Daniel A. Vallero
- Abel Wolman

References

- *Halpin, Daniel W.; Senior, Bolivar A. (2010), Construction Management (4 ed.), Hoboken, NJ: John Wiley & Sons, p. 9, ISBN 9780470447239, retrieved May 16, 2015*
- *Chitkara, K. K. (1998), Construction Project Management, New Delhi: Tata McGraw-Hill Education, p. 4, ISBN 9780074620625, retrieved May 16, 2015*
- Figures from the United Nations' *UN National Accounts Database*. for the countries of the world. Retrieved 1 February 2014. Johnson, Anthony, *Solving Stonehenge: The New Key to an Ancient Enigma*. (Thames & Hudson, 2008) ISBN 978-0-500-05155-9

- *Levathes, Louise (1994). When China Ruled the Seas. Oxford University Press. p. 77. ISBN 978-0-19-511207-8.*
- *Noel, John V. (1988). Knight's Modern Seamanship (18 ed.). John Wiley and Sons. pp. 81–82. ISBN 0-471-28948-5.*
- *Sally Ganchy, Sarah Gancher (2009), Islam and Science, Medicine, and Technology, The Rosen Publishing Group, p. 41, ISBN 1-4358-5066-1*
- Howard R. Turner (1997), *Science in Medieval Islam: An Illustrated Introduction*, p. 181, University of Texas Press, ISBN 0-292-78149-0
- *Tchobanoglous, G.; Burton, F.L. & Stensel, H.D. (2003). Wastewater Engineering (Treatment Disposal Reuse) / Metcalf & Eddy, Inc. (4th ed.). McGraw-Hill Book Company. ISBN 0-07-041878-0.*
- *Turner, D.B. (1994). Workbook of atmospheric dispersion estimates: an introduction to dispersion modeling (2nd ed.). CRC Press. ISBN 1-56670-023-X.*
- *Beychok, M.R. (2005). Fundamentals Of Stack Gas Dispersion (4th ed.). author-published. ISBN 0-9644588-0-2*
- Davis, M. L. and D. A. Cornwell, (2006) *Introduction to environmental engineering* (4th ed.) McGraw-Hill ISBN 978-0072424119
- Compare: *"Construction", Merriam-Webster.com, Merriam-Webster, retrieved 2016-02-16, [...] the act or process of building something (such as a house or road) [...].*
- *"Global construction industry faces growing threat of economic crime". pwc. pwc. Retrieved 16 September 2015.*
- *Swanson, Naomi; Tisdale-Pardi, Julie; MacDonald, Leslie; Tiesman, Hope M. (13 May 2013). "Women's Health at Work". National Institute for Occupational Safety and Health. Retrieved 21 January 2015.*
- *D. Masse; JL. Chotte; E. Scopel (2015). "Ecological engineering for sustainable agriculture in arid and semiarid West African regions". Fiche thématique du CSFD (11): 2.*
- *Diaz, Jesus (2013-01-31). "This Is What the First Lunar Base Could Really Look Like". Gizmodo. Retrieved 2013-02-01.*
- *"Architecture and Engineering Occupations : Occupational Outlook Handbook : U.S. Bureau of Labor Statistics". Bls.gov. 2012-03-29. Retrieved 2013-07-01.*
- *"Funding - Environmental Engineering - US National Science Foundation (NSF)". nsf.gov. Retrieved 2013-07-01.*
- *Wood, Hannah (17 January 2012). "UK Construction Careers, Certifications/Degrees and occupations". TH Services. Retrieved 4 March 2012.*

5

Civil Engineering and Urban Planning

One of the main applications of civil engineering is its vital importance in urban planning and development. Civil engineering is required for the development of any city or a village. This chapter will particularly discuss the topics like planned community, parks, shopping malls, municipal and urban engineering and urban planning. It is one of the most important roles civil engineering plays in our society. Therefore, this chapter is written in such a way that it provides fair information about this aspect of civil engineering to students.

Municipal or Urban Engineering

Municipal or urban engineering applies the tools of science, art and engineering in an urban environment.

Summary

Municipal engineering is concerned with municipal infrastructure. This involves specifying, designing, constructing, and maintaining streets, sidewalks, water supply networks, sewers, street lighting, municipal solid waste management and disposal, storage depots for various bulk materials used for maintenance and public works (salt, sand, etc.), public parks and cycling infrastructure. In the case of underground utility networks, it may also include the civil portion (conduits and access chambers) of the local distribution networks of electrical and telecommunications services. It can also include the optimizing of garbage collection and bus service networks. Some of these disciplines overlap with other civil engineering specialties, however municipal engineering focuses on the coordination of these infrastructure networks and services, as they are often built simultaneously (for a given street or development project), and managed by the same municipal authority.

History

Modern municipal engineering finds its origins in the 19th-century United Kingdom, following the Industrial Revolution and the growth of large industrial cities. The threat to urban populations from epidemics of waterborne diseases such as cholera and typhus lead to the development of a profession devoted to “sanitary science” that later became “municipal engineering”.

A key figure of the so-called "public health movement" was Edwin Chadwick, author of the parliamentary report "The Sanitary Condition of the Labouring Population", published in 1842.

Early British legislation included:

- Burgh Police Act 1833 - powers of paving, lighting, cleansing, watching, supplying with water and improving their communities.
- Municipal Corporations Act 1835
- Public Health Act 1866 – formation of drainage boards
- Public Health Act 1875 known at the time as the Great Public Health Act

This legislation provided local authorities with powers to undertake municipal engineering projects and to appoint borough surveyors (later known as "municipal engineers").

In the U.K, the Association of Municipal Engineers, (subsequently named Institution of Municipal Engineers), was established in 1874 under the encouragement of the Institution of Civil Engineers, to address the issue of the application of sanitary science. By the early 20th century, Municipal Engineering had become a broad discipline embracing many of the responsibilities undertaken by local authorities, including roads, drainage, flood control, coastal engineering, public health, waste management, street cleaning, water supply, sewers, waste water treatment, crematoria, public baths, slum clearance, town planning, public housing, energy supply, parks, leisure facilities, libraries, town halls and other municipal buildings.

In the UK, the development of different strands of knowledge necessary for the management of municipal infrastructure led to the emergence of separate specialised institutions, including:

- For drainage: Chartered Institution of Water and Environmental Management, 1895
- For town planning: Town Planning Institute 1914 ... subsequently becoming the Royal Town Planning Institute
- For street lighting: Association of Public Lighting Engineers, 1934...subsequently becoming the Institution of Lighting Engineers
- For highway engineering: Institution of Highways and Transportation, 1930
- For public housing: Institute of Housing, 1931

In 1984 the Institution of Municipal Engineers merged with the Institution of Civil Engineers.

Since the 1970s, there has been a global trend toward increasing privatisation and outsourcing of municipal engineering services.

In the UK in the 1990s a change in management philosophy brought the demise of the traditional organisational structure of boroughs where the three functions of town clerk, borough treasurer and borough engineer were replaced by an administrative structure with a larger number of specialised departments.

In the late 1990s and early 21st century there was increasing dissatisfaction over what was perceived to be fractured and dysfunctional public services designed along narrow specialties. A more holistic approach to urban engineering began to emerge as an alternative concept. Critics of the specialised approach included the Commission for Architecture and the Built Environment that complained that the specialised approach to management of the public realm focussed too much on the on efficient movement of vehicles rather than the more general interests of local communities.

Professional Practice

In the United Kingdom there is no longer any formal professional qualification in municipal engineering although there are degree courses available in urban engineering.

A professional certificate in Urban Engineering is available from through the Institution of Incorporated Engineers via the Public Realm Information and Advice Network.

The British Institution of Civil Engineers (ICE) caters to practitioners employed in the public sector, private consultancy and academia through its Proceedings Journal Municipal Engineer. The journal, first published in 1873, has a global scope and covers the whole life cycle of municipal services addressing technical, political and community issues. In addition an Expert Panel responds on behalf of ICE to Government consultations and is represented on the International Federation of Municipal Engineering.

International Organization

The International Federation of Municipal Engineering (IFME) is an organisation comprising professional municipal engineers from all round the world. IFME's mission is to connect municipal engineers, public works professionals, public agencies, institutions and businesses around the world in order that they can share a global pool of knowledge and experience. The aim is to foster continued improvement in the quality of public works and wider community services.

The inaugural meeting was held in 1960 at the UNESCO headquarters in Paris. Membership has grown steadily and in 2009 comprised representatives from national associations in: Australia, Canada, Denmark, Estonia, Finland, Italy, Israel, The Netherlands, New Zealand, Norway, Southern Africa (South Africa, Botawana, Namibia & Zimbabwe), Sweden, UK (England, Scotland, Wales & Northern Ireland) and USA. Belgium and San Marino are presently Corresponding Members.

Related Engineering Disciplines

Municipal or urban engineering combines elements of environmental engineering, water resources engineering and transport engineering.

Relationship to Urban Design or Urban Planning

Today, municipal engineering may be confused with urban design or urban planning. Whereas the urbanist or urban planner may design the general layout of streets and public places, the municipal engineer is concerned with the detailed design. For example, in the case of the design of a new street, the urbanist may specify the general layout of the street, including landscaping, surface finishings and urban accessories, but the municipal engineer will prepare the detailed plans and specifications for the roads, sidewalks, municipal services and street lighting.

Site Civil Works

In the case of large buildings or plants, facilities or campuses, site civil works may be required that are similar in scope or type as municipal infrastructure, namely, access roads, parking lots, potable water supply (including fire hydrants), on-site waste water treatment plants, site drainage including sedimentation and retention ponds or basins, etc. In most engineering consulting firms, Structural Engineering and Municipal Infrastructure are typically separate departments. On a large construction project, the civil engineering design will typically be divided into a structural portion, designed by structural engineers and typically focused on the buildings, and "civil" portion, designed by municipal engineers and focused on the site.

Sources

Index to the Proceedings of the Institution of Municipal Engineers, from 1874

The municipal and sanitary engineer's handbook (1883)

Urban Planning

Urban planning is a technical and political process concerned with the development and use of land, protection and use of the environment, public welfare, and the design of the urban environment, including air, water, and the infrastructure passing into and out of urban areas such as transportation, communications, and distribution networks. Urban Planning is also referred to as urban and regional, regional, town, city, rural planning or some combination in various areas worldwide. Urban planning takes many forms and it can share perspectives and practices with urban design.

Urban planning designs settlements, from the smallest towns to the largest cities. Shown here is Hong Kong from Western District overlooking Kowloon, across Victoria Harbour.

Urban planning guides orderly development in urban, suburban and rural areas. Although predominantly concerned with the planning of settlements and communities, urban planning is also responsible for the planning and development of water use and resources, rural and agricultural land, parks and conserving areas of natural environmental significance. Practitioners of urban planning are concerned with research and analysis, strategic thinking, architecture, urban design, public consultation, policy recommendations, implementation and management.

Urban Planners work with the cognate fields of architecture, landscape architecture, civil engineering, and public administration to achieve strategic, policy and sustainability goals. Early urban planners were often members of these cognate fields. Today urban planning is a separate, independent professional discipline. The discipline is the broader category that includes different sub-fields such as land-use planning, zoning, economic development, environmental planning, and transportation planning.

History

There is evidence of urban planning and designed communities dating back to the Mesopotamian, Indus Valley, Minoan, and Egyptian civilizations in the third millennium BCE. Archeologists studying the ruins of cities in these areas find paved streets that were laid out at right angles in a grid pattern. The idea of a planned out urban area evolved as different civilizations adopted it. Beginning in the 8th century BCE, Greek city states were primarily centered on orthogonal (or grid-like) plans. The ancient Romans, inspired by the Greeks, also used orthogonal plans for their cities. City planning in the Roman world was developed for military defense and public convenience. The spread of the Roman Empire subsequently spread the ideas of urban planning. As the Roman Empire declined, these ideas slowly disappeared. However, many cities in Europe still held onto the planned Roman city center. Cities in Europe from the 9th to 14th centuries, often grew organically and sometimes chaotically. But many hundreds of new towns were newly built according to preconceived plans, and many others were enlarged with newly planned extensions. Most of these were realized from the 12th to 14th centuries, with a peak-period at the end of the 13th. From the 15th century on, much more is recorded of urban design and the people that were involved. In this period, theoretical treatises on architecture and urban planning start to appear in which theoretical questions are addressed and designs of towns and cities are described and depicted. During the Enlightenment period, several European rulers ambitiously attempted to redesign capital cities. During the Second French Republic, Baron Georges-Eugène Haussmann, under the direction of Napoleon III, redesigned the city of Paris into a more modern capital, with long, straight, wide boulevards.

Planning and architecture went through a paradigm shift at the turn of the 20th century. The industrialized cities of the 19th century grew at a tremendous rate. The pace and style of this industrial construction was largely dictated by the concerns of private business. The evils of urban life for the working poor were becoming increasingly evident as a matter for public concern. The laissez-faire style of government management of the economy, in fashion for most of the Victorian era, was starting to give way to a New Liberalism that championed intervention on the part of the poor and disadvantaged. Around 1900, theorists began developing urban planning models to mitigate the consequences of the industrial age, by providing citizens, especially factory workers, with healthier environments.

Urban planning started to become professionalized during this time. The Town and Country Planning Association was founded in 1899 and the first academic course on urban planning was offered by the University of Liverpool in 1909. In the 1920s, the ideas of modernism and uniformity began to surface in urban planning, and lasted until the 1970s. Many planners started to believe that the ideas of modernism in urban planning led to higher crime rates and social problems. Urban planners now focus more on individualism and diversity in urban centers.

Theories

Planning Theory is the body of scientific concepts, definitions, behavioral relationships, and assumptions that define the body of knowledge of urban planning. There are eight procedural theories of planning that remain the principal theories of planning procedure today: the rational-comprehensive approach, the incremental approach, the transactive approach, the communicative approach, the advocacy approach, the equity approach, the radical approach, and the humanist or phenomenological approach.

Technical Aspects

Technical Aspects of Urban Planning involve the applying scientific, technical processes, considerations and features that are involved in planning for land use, urban design, natural resources, transportation, and infrastructure. Urban planning includes techniques such as: predicting population growth, zoning, geographic mapping and analysis, analyzing park space, surveying the water supply, identifying transportation patterns, recognizing food supply demands, allocating healthcare and social services, and analyzing the impact of land use.

Urban Planners

An urban planner is a professional who works in the field of urban planning for the purpose of optimizing the effectiveness of a community's land use and infrastructure. They formulate plans for the development and management of urban and suburban areas, typically analyzing land use compatibility as well as economic, environmental and social trends. In developing the plan for a community (whether commercial, residential, agricultural, natural or recreational), urban planners must also consider a wide array of issues such as sustainability, air pollution, traffic congestion, crime, land values, legislation and zoning codes.

The importance of the urban planner is increasing throughout the 21st century, as modern society begins to face issues of increased population growth, climate change and unsustainable development. An urban planner could be considered a green collar professional.

Planned Community

Partizánske in Slovakia– an example of a typical planned industrial city founded in 1938 together with a shoemaking factory in which practically all adult inhabitants of the city were employed.

Abuja, FCT in Nigeria, which was built mainly in the 1980s, was the fastest growing city in the world between 2000 and 2010, with an increase of 139.7%, and still expanding rapidly.

Brasília at night from ISS.

Plan of Fredericia (Denmark) in 1900- the city was founded in 1650.

A planned community, or planned city, is any community that was carefully planned from its inception and is typically constructed in a previously undeveloped area. This contrasts with settlements that evolve in a more *ad hoc* fashion. Land use conflicts are less frequent in planned communities since they are planned carefully. The term new town refers to planned communities of the new towns movement in particular, mainly in the United Kingdom. It was also common in the European colonization of the Americas to build according to a plan either on fresh ground or on the ruins of earlier Amerindian cities.

Planned Capital Cities

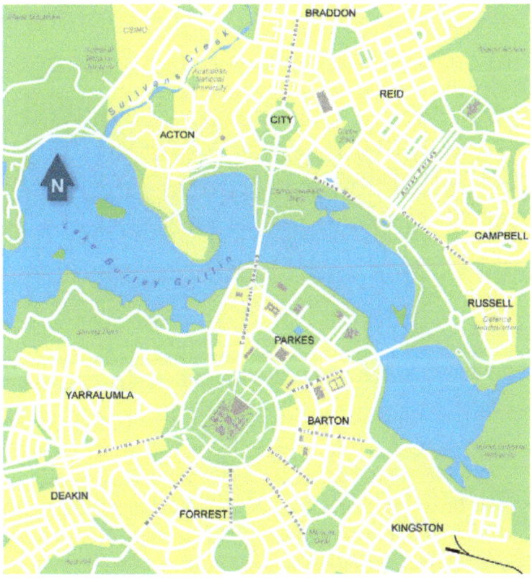

Inner Canberra, Australia.

Several of the world's capital cities are planned cities, including Canberra in Australia, Brasília in Brazil, Belmopan in Belize, New Delhi in India, Valletta in Malta, Abuja in Nigeria, Astana in Kazakhstan, Naypyidaw in Burma, Islamabad in Pakistan and Washington, D.C., in the United States. In Egypt, a new capital city east of Cairo has been proposed. The federal administrative centre of Malaysia, Putrajaya, is also a planned city.

Abu Dhabi (UAE) and some of the recently built cities in the Persian Gulf region are also planned cities, built on the money from the oil boom. Prior to the boom, these were just villages or towns.

Africa

Botswana

The city of Gaborone was planned and constructed in the 1960s.

Equatorial Guinea

In 2012, President Teodoro Obiang decided to move the capital to a new jungle site at Oyala.

Kenya

Konza Technology City is a planned city that is hoped to become a hub of African science and technology upon its completion in 2030.

Nigeria

The capital, Abuja is a planned city and was built mainly in the 1980s. Several other cities are under development to accommodate the rapidly growing population, some of which include: Eko Atlantic City, a planned city of Lagos State being constructed on land reclaimed from the Atlantic Ocean. Upon completion, the new city which is still under development is anticipating 250,000 residents and a daily flow of 150,000 commuters. Centenary City, in the Federal Capital Territory, is another planned smart city under development. The city is designed to become a major tourist attraction to the country. A list of Nigerian cities and neighbourhoods that went through a form of planning are as follows:

- Abuja, one of the most populous planned cities, and the fastest growing city in the world.
- Banana Island, an affluent neighbourhood in Ikoyi, Lagos, Lagos State
- Bonny Island in Rivers State

- Centenary City in the Federal Capital Territory
- Eko Atlantic in Lagos State
- Festac Town in Lagos, Lagos State
    - FESTAC Phase 2, the next phase for Festac has been proposed.
- Ikeja, the capital of Lagos State
- Lekki, a new city in Lagos State
    - Lekki Free Zone, a free trade zone in the Lekki
    - Orange Island, a planned real estate development in Lekki.
- Victoria Island in Lagos, Lagos State

South Africa

A number of cities were set up during the apartheid-era for a variety of ethnic groups. Planned settlements set up for white inhabitants included Welkom, Sasolburg and Secunda. Additionally the majority of settlements in South Africa were planned in their early stages and the original town centres still lie in a grid street fashion. Some settlements were also set up for non whites such as the former homeland capital of Bhisho.

Asia

Burma

Naypyidaw is the capital of Myanmar, also known as Burma. It is administered as the Naypyidaw Union Territory, as per the 2008 Constitution. On 6 November 2005, the administrative capital of Burma was officially moved to a greenfield 3.2 km west of Pyinmana, and approximately 300 km north of Yangon (Rangoon), the previous capital. The capital's official name was announced on 27 March 2006, Burmese Armed Forces Day. Much of the city was still under construction as late as 2012. As of 2009, the population was 925,000, which makes it Burma's third largest city, after Yangon and Mandalay.

Mainland China

Many ancient cities in China, especially those on the North China Plain, were carefully designed according to the fengshui theory, featuring square or rectangular city walls, rectilinear road grid, and symmetrical layout. Famous examples are Chang'an in Tang dynasty and Beijing.

An exception to that is an ancient town in Tekes County, Ili Kazakh Autonomous Prefecture, Xinjiang, with a shape of a ba gua.

In modern China, many special economic zones are developed from the sketch, for example, Pudong, a new district of Shanghai.

Hong Kong

Part of Tuen Mun New Town, developed from the 1970s onward

The area of Hong Kong is very mountainous and many places in the New Territories have limited access to roads. Hong Kong started developing new towns in the 1950s, to accommodate rapidly growing populations. In the early days the term "satellite towns" was used. The very first new towns included Tsuen Wan and Kwun Tong. Wah Fu Estate was built in a remote corner on Hong Kong Island, with similar concepts in a smaller scale.

In the late 1960s and the 1970s, another stage of new town developments was launched. Nine new towns have been developed to date. Land use is carefully planned and development provides plenty of room for public housing projects. Rail transport is usually available at a later stage. The first towns are Sha Tin, Tsuen Wan, Tuen Mun and Tseung Kwan O. Tuen Mun was intended to be self-reliant, but was not successful and turned into a bedroom community like the other new towns. More recent developments are Tin Shui Wai and North Lantau. The government also plans to build such towns in Hung Shui Kiu, Ping Che-Ta Kwu Ling, Fanling North and Kwu Tung North. At present, there are a total of nine new towns:

- Tsuen Wan New Town
- Sha Tin New Town

- Tuen Mun New Town
- Tai Po New Town
- Fanling-Sheung Shui New Town
- Yuen Long New Town
- Tseung Kwan O New Town
- Tin Shui Wai New Town
- North Lantau New Town

Indonesia

Bandung laid as a well-planned city, set as the new capital of the Dutch East Indies back in 1920s.

- Batavia (predecessor of modern Jakarta) was a planned city, modeled after Dutch 17th century coastal city architecture. First, in 17th century as a planned fortified city, crisscrossed with Dutch-style canals dug in regular grid. The city served as the administrative center of Dutch East India Company.
- In early 19th century, the Dutch colonial authority moved their administrative center from dilapidated and unhealthy port town of Old Batavia (now Kota area) several kilometres south to Weltevreden area (now Central Jakarta). Old Batavia and Weltevreden were connected by the Molenvliet Canal and a road (now Gajah Mada Road) that ran alongside the waterway. It was a well-planned community around the Koningsplein, the Waterlooplein and Rijswijk (Jalan Veteran). The area, then known as Weltevreden, which include the Koningsplein, Rijswijk, Noordwijk, Tanah Abang, Kebon Sirih, and Prapatan became a popular residential, entertainment and commercial district for the European colonial elite.

- Menteng, today a subdistric in Central Jakarta, was first built as a well-planned community. An urban design developed in the 1910s set the area to become a residential area for Dutch people and high officials. At the time of its development, the area was the first planned garden suburb in colonial Batavia. Supported by easy access to service centers and nearby to the central business district, this area has become one of the most expensive areas for residential real estate in modern Jakarta.
- In early 20th century, Bandung was planned by Dutch East Indies government as a new capital city to replace Batavia. The idea was to separate the busy trading port or the commercial center (Batavia) from the new administrative and political center (Bandung). By 1920s the plan to transfer the capital to Bandung was underway. As the city began to laid the masterplan of a well-planned new city, grid of streets and avenues were laid, and numbers of government buildings were constructed, such as Gedung Sate that was planned as the colonial administrative center of Dutch East Indies. The plan however, was failed due to the Great Depression and the outbreak of the Second World War.
- Since Palangkaraya was established as the capital of Central Kalimantan province in 1957, the first president of Indonesia, Sukarno, outlined a plan to develop Palangkaraya as the future capital of Indonesia. Palangkaraya is far larger in area than Jakarta and safe from the danger of earthquakes and volcanoes, common on the island of Java.
- In late 1950s to first half of 1960s, Sukarno, Indonesia's first president, laid a masterplan to built Jakarta as the planned national capital of the Republic of Indonesia. He filled Jakarta with numbers of monuments and statues. Numbers of monumental projects were conceived, planned and initiated during his administration, including Monumen Nasional, Istiqlal mosque, DPR/MPR Building, and Gelora Bung Karno stadium. Sukarno also filled Jakarta with nationalistic monuments and statues, including Selamat Datang Monument, Pemuda Monument at Senayan, Dirgantara Monument at Pancoran, and Irian Jaya Liberation Monument at Lapangan Banteng. Although many of this projects were completed later in his successor era (Suharto administration), Sukarno is credited for shaping Jakarta's monuments and landmarks. He desired Jakarta to be the beacon of a powerful new nation.
- Because of Jakarta's environmental degradation and overpopulation problems, in 2010s there has been an idea to build a new proposed capital city to replace Jakarta.

Iran

In the period of the Persian Safavid Empire, Isfahan, the Persian capital, was built according to a planned scheme, consisting of a long boulevard and planned housing and green areas around it.

In modern-day Iran more than 20 planned cities have been developed or are under construction, mostly around Iran's main metropolitan areas such as Tehran, Isfahan, Shiraz and Tabriz. Some of these new cities are built for special purposes such as:

- Pardis, which is built as a scientific city.
- Poulad-Shahr, which is an industrial city built for the housing of Isfahan's steel industry workers.
- Shirin Shahr which is to provide housing for the sugar industry personnel.
- Tehranpars which was built to house Tehran's additional population.
- Shahrak-e Gharb, built as a massive project of modern apartment buildings.
- Parand which is intended to provide residences for the staff of Imam Khomeini International Airport.
- Shushtar New Town which was built to provide housing for the employees of a sugar cane processing plant.

576,000 people were planned to be settled in Iran's new towns by the year 2005.

Israel

A planned community in the Negev

According to politics of country settlement a number of planned cities were created in peripheral regions. Those cities also known as *Development Towns*. The most successful is Ashdod with more than 200,000 inhabitants, a port and developed infrastructure. Other cities that were developed following Israel's lineation plan are Shoham, Karmiel and Arad. Modi'in-Maccabim-Re'ut has been another of the country's most successful planned cities. Construction began in 1994 and it now has a population of over 80,000. Modi'in also rates higher in terms of average salary and graduation rates than the national average. It was designed and planned by Israeli architect Moshe Safdie. Many Israeli settlements follow this model, including towns like Modi'in Illit and Betar Illit.

Japan

Kyoto was built on a grid system, starting in 794.

The city of Kyoto was developed as a planned city in 794 as a new imperial capital (then called Heian-kyō), built on a grid layout modeled after the Tang dynasty capital of Chang'an (modern day Xi'an), and remained the capital for over a millennium. The grid layout remains, reflected in major east-west streets being numbered, such as 4th street (四条 *shi-jō*?). In modern times, Sapporo was built from 1868, following an American grid plan, and is today the fifth-largest city in Japan. Both these cities have regular addressing systems (following the grid) unlike the usual subdivision-based Japanese addressing system.

Borrowing from the New Town movement in the United Kingdom, some 30 new towns have been built all over Japan. Most of these constructions were initiated during the period of rapid economic growth in the 1960s, but construction continued into the 1980s. Most of them are located near Tokyo and the Kansai region. Some towns, (Senri New Town, Tama New Town) do not provide much employment, and many of the residents commute to the nearby city. These towns fostered the infamous congestion of commuter trains (although as the metropolitan areas have grown, this commute has become relatively short in comparison to commutes from the new urban fringe).

Other New Towns act as industrial/academic agglomerations (*sangyo-shuseki*) (Tsukuba Science City, Kashima Port Town). These areas attempt to create an all-inclusive environment for daily living, in accordance with Uzō Nishiyama's "life-spheres" principle.

Japan has also developed the concept of new towns to what Manuel Castells and Sir Peter Hall call technopolis. The technopolis program of the 1980s has precedents in the

New Industrial Cities Act of the 1960s. These cities are largely modeled after Tsukuba Academic New Town (Tsukuba Science City) in that they attempt to agglomerate high-tech resources together in a campus-like environment.

In the past, the Japanese government had proposed relocating the capital to a planned city, but this plan was cancelled.

Overall, Japan's New Town program consists of many diverse projects, most of which focus on a primary function, but also aspire to create an all-inclusive urban environment. Japan's New Town program is heavily informed by the Anglo-American Garden City tradition, American neighborhood design, as well as Soviet strategies of industrial development.

In 2002 Prime Minister Junichiro Koizumi announced the end of new town construction, although the new towns continue to receive government funding and redevelopment.

Sources:

- Ministry of Construction, Japan International Cooperation Agency, City Bureau. 1975? City Planning in Japan.
- Hein, Carola. 2003. "Visionary Plans and Planners: Japanese Traditions and Western Influences" in Japanese Capitals in Historical
- Perspective, Nicholas Fiévé and Paul Waley, eds. New York: RoutledgeCurzon, 309-43.
- Scott, W. Stephen. 2006. Just Housing? Evidence of Garden City Principles in a Postwar Japanese New Town. Undergraduate diss. New College of Florida.

Malaysia

- Kulim Hi-Tech Park
- See entries for Nusajaya and Iskandar Malaysia

Palestinian Territories

- Rawabi

Philippines

Quezon City was the planned city of President Manuel L. Quezon, who had earlier proposed a new city to be built on land northeast of the City of Manila. Carefully-planned districts include Santa Mesa Heights (part of the original Burnham Plan for Manila), the Diliman Estate (includes the University of the Philippines), New Manila, the Cubao Commercial District, South Triangle, Housing Projects 1 (Roxas District), 2 and 3 (Quirino District), 4, 5 (Kamias-Kamuning District), 6, 7, and 8.

President Elpidio Quirino proclaimed Quezon City as the national capital on July 17, 1948, with President Ferdinand Marcos restoring Manila as the capital on June 24, 1976. He then created a metropolitan area called Metro Manila, which remains congested today due to failed execution of the Quezon City plan as well as the Burnham Plan.

Other planned cities (in order of foundation):

- Baguio
- Trece Martires, Cavite
- Island Garden City of Samal
- Palayan, Nueva Ecija

Saudi Arabia

King Abdullah Economic City, a future planned city along the Red Sea located in Saudi Arabia.

In 1975, Jubail Industrial City, also known as Jubail, was designated as a new industrial city by the Saudi government. It provides 50% of the country's drinking water through desalination of the water from the Persian Gulf.

Singapore

The new town planning concept was introduced into Singapore with the building of the first New Town, Queenstown, from July 1952 to 1973 by the country's public housing authority, the Housing and Development Board. Today, the vast majority of the approximately 11,000 public housing buildings are organised into 22 new towns across the country.

Each new town is designed to be completely self-sustainable. Helmed by a hierarchy of commercial developments, ranging from a town centre to precinct-level outlets, there is no need to venture out of town to meet the most common needs of residences. Employment can be found in industrial estates located within several towns. Educational, health care, and recreational needs are also taken care of with the provision of schools, hospitals, parks, sports complexes, and so on.

Singapore's expertise in successful new town design was internationally recognised when the Building and Social Housing Foundation (BSHF) of the United Nations awarded the World Habitat Award to Tampines New Town, which was selected as a representative of Singapore's new towns, on 5 October 1992.

South Asia

Ancient History

A sophisticated and technologically advanced urban culture is evident in the mature phase of Indus Valley Civilization which thrived in present-day Pakistan and north western India from around 3300 BC. The quality of municipal city planning suggests knowledge of urban planning and efficient municipal governments which placed a high priority on hygiene. The streets of major cities in present-day Pakistan such as Mohenjo-daro and Harappa, the world's earliest planned cities, were laid out in a perfect grid pattern comparable to that of present-day New York City. The houses were protected from noise, odours, and thieves.

As seen in the ancient sites of Harappa and Mohenjo-daro in Pakistan and western border of India, this urban plan included the world's first urban sanitation systems. Within the city, individual homes or groups of homes obtained water from wells. From a room that appears to have been set aside for bathing, waste water was directed to covered drains, which lined the major streets. Houses opened only to inner courtyards and smaller lanes.

The ancient Indus systems of sewage and drainage that were developed and used in cities throughout the Indus Valley were far more advanced than any found in contemporary urban sites in the Middle East and even more efficient than those in some areas of modern South Asia today. The advanced architecture of the Harappans is shown by their impressive dockyards, granaries, warehouses, brick platforms, and protective walls.

Medieval History

A number of medieval Indian cities were planned including:

- Ahmedabad in Gujarat. Established by Sultan Ahmad Shah in 1411.
- Jaipur in Rajasthan. Established in 1727 by Maharaja Sawai Jai Singh II.It is the capital of Rajasthan state of India.
- Udaipur in Rajasthan. It was the historic capital of the former kingdom of Mewar.
- Madurai, in the state of Tamil Nadu. It was the capital of the erstwhile Pandyan kingdom and is noted for its lotus-like symmetry.

- Fatehpur Sikri in Agra. Its planning was done by the Mughal emperor Akbar the Great.
- Vijayanagar in Karnataka, the capital of the erstwhile Vijayanagara Empire.
- Hampi in Karnataka, which was the former capital of the Vijayanagara Empire prior to the city of Vijayanagar.

Modern History

Bangladesh

All except the last one are satellites of Dhaka, Bangladesh's capital.

- Dhanmondi
- Gulshan
- Banani
- Baridhara
- Uttara Model Town
- Banasree
- Bashundhara
- Niketon
- Motijheel & Dilkusha (central business district)
- Tejgaon (an industrial area)
- Kawran Bazar (a commercial hub)
- Khilgaon
- Mirpur
- Purbachal New Town (will be the capital of Bangladesh and now is under-construction)

India

The period following independence saw India being defined into smaller geographical regions. New states such as Gujarat were formed with planned capital cities. The major planned cities of India include:

- Navi Mumbai, World's largest planned city. Contains a population of 2.6 million.
- Gurgaon
- Manesar (Gurgaon)
- New Gurgaon
- Rajarhat (New Town, Kolkata), a very beautiful planned township near Kolkata
- New Delhi
- NOIDA
- Chandigarh India's first planned city
- Durgapur India's second planned city
- Visakhapatnam
- Gandhinagar
- Patiala
- Panchkula
- Mohali
- Greater Noida
- Lavasa
- Udaipur
- Auroville
- Bhubaneshwar, the capital of Odisha
- Rourkela
- Bokaro Steel City
- Dispur
- Dhule, Maharashtra, a city planned by the renowned architect of India, Sir Vishveshwarayya.
- Naya Raipur, the upcoming capital of Chhattisgarh
- Amaravati, a new city planned to become the capital of Andhra Pradesh

- Bidhannagar, Kolkata(Salt Lake City), planned township near Kolkata
- Kalyani
- Jamshedpur, planned township in Jharkhand
- Bhilai, planned township in Chhattisgarh
- Pimpri-Chinchwad
- Solapur, planned drainage city

Pakistan

- The most notable planned city is the capital Islamabad, whose first foundations were laid during the 1950s. Islamabad was one of the most successfully built and planned cities in the world.
- Jauharabad- a planned town situated in Khushab District, Punjab, Pakistan.
- Sargodha Established by British for Air Defence
- Faisalabad Established by British and later Ayub Khan for Industry
- Gwadar Established by government as the southern leg for China–Pakistan Economic Corridor
- Sahiwal
- LDA City
- DCK
- Bahria Town's in Karachi, Lahore and Rawalpindi are privately developed towns.
- Clifton
- DHA Karachi
- DHA Lahore

South Korea

New Songdo City is a planned international business centre to be developed on 6 square kilometres of reclaimed land along Incheon's waterfront, 65 kilometres west of Seoul and connected to Incheon International Airport by a 10 kilometre highway bridge. This 10-year development project is estimated to cost in excess of $40 billion, making it the largest private development project ever undertaken anywhere in the world.

Gwanggyo newtown is located 25 km south away from Seoul in Suwon city and Youngin city, Gyeonggi province. Gwanggyo newtown area 11 square kilometers was designated in 2004 by Gyeonggi Province, Suwon city, Youngin city, and Gyeonggi Development Corporation(GICO). It will accommodate more than 31,000 households. Gwanggyo newtown was not only for the housing supply but also for several regional goals such as provincial office movement, convention center building, and creating economic growth core in Gyeonggi provincial area. Its infrastructure is scheduled to be constructed by 2012.

Since the 1990s, several planned communities were built in the Seoul Metropolitan Area to alleviate housing demands in Seoul. They include:

- Several ongoing developments in Hwaseong, including Bongdam, Dongtan, and Hyangnam
- Bundang, Seongnam City
- Ilsan, Goyang City
- Pangyo, Seongnam City

Since 2007 Sejong City was planned as the new capital but is becoming a research hub instead with many national research institutes moving there between 2013 and 2015. It has 0.8 million of planned population, which is the largest in all of the new town development plan ever.

Taiwan

- Zhongxing Xincun, Nantou County
- Linkou District, New Taipei—Guishan District, Taoyuan City (Linkou New Town, 林口新市鎮)
- Danhai, Tamsui District, New Taipei (Danhai New Town, 淡海新市鎮)
- I-da (I-da World, 義大)

United Arab Emirates

- Capital city of Abu Dhabi, is a planned city to some extent
- Certain new parts of Dubai, are planned
- Masdar City, conceived of as a mixed purpose residential and commercial area

Europe

History

New settlements were planned in Europe at least since Greek antiquity. The Greeks built new colonial cities around the Mediterranean. The ancient Romans also founded many new colonial towns through their empire. There are, however, also traces of planned settlements of non-Roman origin in pre-historic northern Europe. Most planned settlements of medieval Europe were created in the period of about the 12th to 14th centuries. All kinds of landlords, from the highest to the lowest rank, tried to found new villages and towns on their estates, in order to gain economical, political or military power. The settlers generally were attracted by fiscal, economical and juridical advantages granted by the founding lord, or were forced to move from elsewhere from his estates. Most of the new towns were to remain rather small (as for instance the bastides of southwestern France), but some of them became important cities, such as Cardiff, Leeds, 's-Hertogenbosch, Montauban, Bilbao, Malmö, Lübeck, Munich, Berlin, Bern, Klagenfurt, Alessandria, Warsaw and Sarajevo.

Roman Empire

The Romans built a large number of towns throughout their empire, often as colonies for the settlement of citizens or veterans. These were generally characterised by a grid of streets and a planned water-supply; and many modern European towns of originally Roman foundation still retain part of the original street-grid. The most impressive Roman planned town was the city of Constantinople from around the 4th century. Roman Emperor Constantine the Great chose the site for the new metropolis and began construction. His plans quickly fell into place. The modern city (now known as Istanbul) has changed much since then, but it must be remembered that the city did not develop due to simple human migrational patterns nor pure military advantage. Constantine wanted a city to mark his magnificence and Constantinople fulfilled the desire.

Belgium

Louvain-la-Neuve, built for the Université Catholique de Louvain.

Bosnia and Herzegovina

- Slobomir is a new town in Bosnia and Herzegovina and its name means: "the city of freedom and peace". It is located on the Drina river near Bijeljina. It was founded by Slobodan Pavlović, a Bosnian philanthropist. It aims to be one of the major cities of post-war Bosnia and Herzegovina. In fact, the city will be located in two countries, Bosnia and Herzegovina and Serbia, although majority of it will be in Bosnia and Herzegovina. The city is named after its founder, Slobodan Pavlović, and his wife, Mira.

- Andrićgrad is town under construction by the famous Serbian director Emir Kusturica, and will be located in Višegrad, Republika Srpska.

Bulgaria

The cities of Stara Zagora and Kazanlak, in central Bulgaria, were rebuilt as planned cities after they were burnt to the ground in the 1877-1878 Russia-Turkey War. Also the city of Dimitrovgrad in south Bulgaria, that was planned as a key industrial and infrastructure center.

Croatia

Červar-Porat is a resort town in western Croatia, located on the east coast of the Adriatic Sea on the Červar lagoon. It was built as planned town in the 1970s, although the area was inhabited in Roman times. During the War of Independence it was used as a camp for refugees from Bosnia and Herzegovina and Vukovar. It was planned to house 6500 people.

The capital of Zagreb underwent major expansion during the 60's. By that time, the city's official boundary was the river Sava, since nothing was built over it. After a flood in the 60's, many residents were moved and some other districts were created for the residents, such as Dubrava, which was the interconnection between the Zagreb's old part and Sesvete. During the 60's and 70's, a planned part of Zagreb, Novi Zagreb (*New Zagreb*), was constructed, which lied on the other, uninhabited part of the river Sava, and is now one of major districts consisting of purely residential buildings and blocks. It is still under expansion and some new landmarks were built in it, the most famous one is the recent one, Arena Zagreb, built in 2008.

Czechoslovakia and Czech Republic

Poruba and Havířov were established in the 1950s as new satellite residential towns for workers of coal-mining, steel-mill and other heavy-industry complex in the Ostrava region.

Prague was extended by large housing estates - "new towns" in the 1970s and 1980s: Severní Město (Northern Town), Jižní Město (Southern Town), Jihozápadní Město (South-Western Town) were the largest, with population around 100.000 each. Their remote position to the city centre was compensated for by underground lines constructed usually a decade after the completion of the housing projects.

Denmark

Fredericia was designed as a combination of town and military fortress following the devastation caused by the Thirty Years' War. A more recent example is Ørestad, planned and built to strengthen development in the Copenhagen/Malmö region. The suburb Albertslund was also built from scratch in the 1970s, merging the two villages Vridsløselille and Herstedvester.

Finland

The city of Helsinki, previously a town of 5,000 inhabitants, was made the capital of the new Grand Duchy of Finland in 1812 by decree of Alexander I, Emperor of Russia. The city center was rebuilt with the lead of the German architect Carl Ludvig Engel.

However, the last city in Finland that was ordered to be built on a previously completely uninhabited land was Raahe, founded by governor general Per Brahe the Younger in 1649.

Finland also has various "ekokylä" communities or "ecological villages". For example, Tapiola is a post-war garden city on the edge of Espoo.

The city of Vaasa was rebuilt about seven kilometers northwest of its original location in 1862, after a fire which destroyed the city in 1852. The new town was planned by Carl Axel Setterberg. The disastrous consequences of the fire were considered as the design included five broad avenues which divided the town into sections and each block was divided by alleys.

Hamina is an old Finnish Eastern trade capital, founded during the Swedish reign. The star-shaped fortress and the circular town plan are based on an Italian Renaissance fortress concept from the 16th century. Fortress towns like this are quite rare, another example is Palmanova in Italy.

France

Many new cities, called bastides, were founded from the 12th to 14th centuries in southwestern France, where the Hundred Years War took place, in order to replace destroyed cities and organize defence and growth. Among those, Monpazier, Beaumont, and Villeréal are good examples.

In 1517, the construction of Le Havre was ordered by Francis I of France as a new port. It was completely destroyed during the Second World War and was entirely rebuilt in a modernist style, during the Trente Glorieuses, the thirty-year period from 1945-1975.

Cardinal Richelieu founded the small Baroque town of Richelieu, which remains largely unchanged.

A program of new towns (French *ville nouvelle*) was developed in the mid-1960s to try to control the expansion of cities. Nine *villes nouvelles* were created.

- Near Paris: Cergy-Pontoise, Marne-la-Vallée, Sénart (former Melun-Sénart), Évry, Saint-Quentin-en-Yvelines

- Near Lille: Villeneuve d'Ascq (Former Lille-Est)
- Near Lyon: L'Isle-d'Abeau
- Near Marseille: Rives de l'Etang de Berre
- Near Rouen: Val-de-Reuil
- Near Grenoble: L'Isle-d'Abeau

La Défense, in the greater Paris area, could also be considered a planned town, though it was not built all at once but in successive stages beginning in the 1950s.

Germany

Planned cities in Germany are:

- Bayreuth: an example of a medieval new city
- Berlin - Friedrichstadt
- Eisenhüttenstadt: the "first socialist town" in Germany
- Freudenstadt: the roads follow the layout of the Nine Men's Morris game
- Halle-Neustadt: a "Stadtteil" or borough in Halle, Saxony-Anhalt
- Karlsruhe: the roads follow the layout of a hand-held fan with the castle being at the juncture
- Ludwigsburg: planned new capital for the duke of Württemberg
- Mannheim Quadratestadt: squares named like ranks and files on a chessboard
- Munich Maxvorstadt: the first planned city expansion of Munich was realized from 1805 to 1810 according to a raster
- Neustrelitz: founded in 1733 with streets spreading from an octagonal market place
- Putbus: built around a circular centre with radially aligned streets
- Wolfsburg: founded in 1938 to host the factories for the newly built Volkswagen

Welthauptstadt Germania was the projected renewal of Berlin as a planned city, although only a small portion was constructed between 1937 and 1943.

Greece

Planned cities in Greece are:

- Lakki: planned and built in the 1930s by the Italians to house members of its military stationed on the island of Leros; its initial name was Portolago

- Nea Alikarnassos: founded in 1925 to house Greek refugees
- Orestiada: founded in 1922 to house Greek refugees
- Paralia Distomou: planned and built to house workers of Aluminium of Greece
- Sparta: planned and built in 1834 after a decree issued by Otto of Greece

Hungary

All Hungarian planned cities were built in the second half of the 20th century when a program of rapid industrialization was implemented by the communist government.

- The Akadémiaújtelep area of Budapest was designed in a unique geometrical fashion.
- Dunaújváros, built next to the existing village Dunapentele to provide housing for workers of a large steel factory complex. Once named after Stalin, the city maintains its importance in heavy industry even after the recession following the end of Communist era.
- Tiszaújváros, built next to the existing village Tiszaszederkény and was named after Lenin for decades. A significant chemical factory was built simultaneously.
- Kazincbarcika, created from the villages Sajókazinc, Barcika and Berente (the latter has become independent since then) in a mining area. The city and its population grew fast after the founding of a factory.
- Tatabánya, created from four already existing villages was developed into a mining town and industrial centre and shortly after its elevation to town status became the county seat of its county, a status it still maintains despite the presence of historically more significant towns in the area.
- Beloiannisz (although not a town, only a village) was planned and built in the 1950s to provide home for Greek refugees of the Civil War.

Ireland

In the Republic of Ireland, as not in the United Kingdom, the term "new town" is often used to refer to planned towns built after World War II which were discussed as early as 1941. The term "new town" in Ireland was also used for some earlier developments, notably during the Georgian era. Part of Limerick city was built in a planned fashion as "Newtown Pery".

In 1961 the first new town of Shannon was commenced and a target of 6,000 inhabitants was set. This has since been exceeded. Shannon is of some regional importance today as an economic centre (with the Shannon Free Zone and Shannon International Airport), but until recently failed to expand in population as anticipated. Since the late 1990s, and particularly in the early 2000s, the population has been expanding at a much faster rate, with town rejuvenation, new retail and entertainment facilities and many new housing developments.

It was not until 1967 that the Wright Report planned four towns in County Dublin. These were Blanchardstown, Clondalkin, Lucan and Tallaght but in actuality this was reduced to Blanchardstown, Lucan-Clondalkin and Tallaght. Each of these towns has approximately 50,000 inhabitants today.

The most recent new town in Ireland is Adamstown in County Dublin. Building commenced in 2005 and it is anticipated that occupation will commence late in 2006 with the main development of 10,500 units being completed within a ten-year timescale.

Italy

In the past centuries several new towns have been planned in Italy. One of the most famous is Pienza, close to Siena, a Renaissance city, also called The Ideal Town or Utopia Town. Between 1459 and 1462 the most famous architects of Italy worked there for the Pope Pius II and built the city centre of the small town.

Another example of renaissance planned cities is the walled star city of Palmanova. It is a derivative of ideal circular cities, notable Filarete's imaginary Sforzinda.

In early 20th century, during the fascist government of Benito Mussolini, many new cities were founded, the most prominent being Littoria (renamed Latina after the fall of the Fascism). The city was inaugurated on December 18, 1932. Littoria was populated with immigrants coming from Northern Italy, mainly from Friuli and Veneto.

The great Sicilian earthquake of 1693 forced the complete rebuilding on new plans of many towns.

Other well known new cities are located close to Milan in the metropolitan area. Crespi d'Adda, a few kilometres east of Milan along the Adda River, was settled by the Crespi family. It was the first Ideal Worker's City in Italy, built close to the cotton factory. Today Crespi d'Adda is part of the Unesco World Heritage List. Cusano Milanino was settled in the first years of the 20th century in the formerly small town of Cusano. It was built as a new green city, rich in parks, villas, large boulevards and called Milanino (Little Milan). In the 1970s in the eastern metropolitan area of Milan a new city was built by Silvio Berlusconi. It is called Milano Due. It is a garden city designed for families of the upper middle class, with peculiarity of having pedestrian paths completely free of traffic. In the 1980s another two similar cities were built by Berlusconi, Milano 3 and Milano Visconti. Each of them has around 12,000 inhabitants.

Lithuania

In 1961 Elektrėnai was established as planned city for workers in Elektrėnai Power Plant and in 1975 Visaginas was established as planned city for workers in Ignalina Nuclear Power Plant.

Macedonia

The graphical scheme of the Detailed Urbanist Plan for a settlement within the Municipality of Aerodrom within the City of Skopje, Republic of Macedonia.

The Municipality of Aerodrom within the City of Skopje is a planned community.

Malta

- The fortified cities of Senglea and Valletta were both built on a grid plan by the Knights of Malta in the 16th century.
- The town of Paola, also known as *Raħal Ġdid* (New Town), is built on a grid plan by the Grand Master Antoine de Paule.
- The towns of San Ġwann and Santa Luċija were built as planned cities in the 1960s and 1970s

Netherlands

One province of the Netherlands, Flevoland (pop. 370,000 (2006)), was reclaimed from the Zuiderzee (Southern Sea). After a flood in 1916, it was decided that the Zuiderzee, an inland sea within the Netherlands, would be closed and reclaimed. In 1932, a causeway (the Afsluitdijk) was completed, which closed off the sea completely. The Zuiderzee was subsequently called IJsselmeer (IJssel-lake) and its previously salty water became fresh.

The first part of the new lake that was reclaimed was the Noordoostpolder (Northeast polder). This new land included, among others, the former island of Urk and it was included with the province of Overijssel. After this, other parts were also reclaimed: the eastern part in 1957 (Oost-Flevoland) and the southern part (Zuid-Flevoland) in 1968. The municipalities on the three parts voted to become a separate province, which happened in 1986. The capital of Flevoland is Lelystad, but the biggest city is Almere (pop. 183,500 in February 2008). Apart from these two larger cities, several 'New Villages' were built. In the Noordoostpolder the central town of Emmeloord is surrounded by ten villages, all on cycling distance from Emmeloord since that was the most popular way of transport in the 1940s (and it's still very popular). Most noteworthy of these villages is Nagele which was designed by famous modern architects of the time, Gerrit Rietveld, Aldo van Eyck, Willem Wissing and Jaap Bakema among them. The other villages were built in a more traditional/vernacular style. In the more recent Flevoland-polders four more 'New Villages' were built. Initially more villages were planned, but the introduction of cars made fewer but larger villages possible.

New towns outside Flevoland are Hoofddorp and IJmuiden near Amsterdam, Hellevoetsluis and Spijkenisse near Rotterdam and the navy port Den Helder.

The cities of Almere, Capelle aan den IJssel, Haarlemmermeer (also a reclaimed polder, 19th century), Nieuwegein, Purmerend and Zoetermeer are members of the European New Town Platform.

Norway

- Oslo: After a great fire in 1624, it was decided by the then King Christian IV that the city would be moved behind the Akershus fortress. The new town, named Christiania, was laid out in a grid and is now the downtown area known as "Kvadraturen" (*the Quadrature*). The original town of Oslo was later incorporated into Christiania, and is now a neighborhood in eastern Oslo; Gamlebyen or "The Old City".
- The city of Kristiansand was formally founded in 1641 by King Christian IV. The city was granted all trade privileges on the southern coast of Norway, denying all other towns to trade with foreign states. As Oslo/Christiania before it, the city was behind a fortress, with a grid system allowing cannons to fire towards the two ports of the city and the river on the eastern end.

Poland

Four cities stand out as examples of planned communities in Poland: Zamość, Gdynia, Tychy and Nowa Huta. Their very diverse layouts are the result of the different aesthetics that were held as ideal during the development of each of these planned communities. Planned cities in Poland have a long history and fall primarily into three time periods during which planned towns developed in Poland and its neighbors that once comprised the Polish-Lithuanian Commonwealth. These are the Nobleman's Republic (16th to 18th centuries), the interwar period (1918–1939) and Socialist Realism (1944–1956).

The Nobleman's Republic of the Polish-Lithuanian Commonwealth

The extreme opulence that Poland's nobility enjoyed during the Renaissance left Poland's elites with not only obscene amounts of money to spend, but also motivated them to find new ways to invest their hefty fortunes out of the grasp of the Royal Treasury. Jan Zamoyski founded the city of Zamość in order to circumvent royal tariffs and duties while also serving as the capital for his mini-state. Zamość was planned by the renowned Paduan architect Bernardo Morando and modeled on Renaissance theories of the 'ideal city'. Realizing the importance of trade, Zamoyski issued special location charters for representatives of peoples traditionally engaged in trade, i.e. to Greeks, Armenians and Sephardic Jews and secured exemptions on taxes, customs duties and tolls, which contributed to its fast development. Zamoyski's success with Zamość spawned numerous other Polish nobles to found their own "private" cities such as Białystok and many of these towns survive today, while Zamość was added to the UN World Heritage list in 1992 and is today considered one of the most precious urban complexes in Europe and in the world.

Interwar Period

The preeminent example of a planned community in interwar Poland is Gdynia. After World War I when Poland regained its independence it lacked a commercial seaport (*De iure* Poles could use Gdańsk, which was the main port of the country before the War and is again today, but *de facto* the Germans residing in the city made it almost impossible for them), making it necessary to build one from scratch. The extensive and modern seaport facilities in Gdynia, the most modern and extensive port facilities in Europe at the time, became Poland's central port on the Baltic Sea. In the shadow of the port, the city took shape mirroring in its scope the rapid development of 19th-century Chicago, growing from a small fishing village of 1,300 in 1921 into a full blown city with a population over 126,000 less than 20 years later. The Central Business District that developed in Gdynia is a showcase of Art Deco and Modernist architectural styles and predominate much of the cityscape. There are also villas, particularly in the city's villa districts such as Kamienna Góra where Historicism inspired Neo-Renaissance and Neo-Baroque architecture.

Socialist Realism

After the destruction of most Polish cities in World War II, the Communist regime that took power in Poland sought to bring about architecture that was in line with its vision of society. Thus urban complexes arose that reflected the ideals of socialist realism. This can be seen in districts of Polish cities such as Warsaw's MDM. The City of Nowa Huta (now a district of Kraków) and Tychy were built as the epitome of the proletarian future of Poland.

Portugal

Vila Real de Santo António was built after the 1755 Lisbon earthquake, on the same model that was used for rebuilding Lisbon, Portugal's capital city (also destroyed in the earthquake), and on a similar orthogonal plan.

Romania

The cities of Brăila, Giurgiu and Turnu Severin were rebuilt, according to new plans, in the first part of the 19th century and the cities of Alexandria and Călăraşi were built completely new the same time. The city of Victoria, located in the Braşov County, was built by the communist government in the beginning of the second half of the 20th century.

Russia

- Saint Petersburg was built by Peter the Great as a planned capital city starting in 1703.
- Magnitogorsk is an example of a planned industrial city based on Stalin's 1930s five-year plans.
- The Avtozavodsky district of Tolyatti is a planned industrial city of Soviet post-war modernism.

Serbia

Novi Beograd, meaning *New Belgrade* in Serbian, is a municipality of the city of Belgrade, built on a previously undeveloped area on the left bank of the Sava river. The first development began in 1947, the municipality has since expanded significantly and become the fastest developing region in Serbia.

Drvengrad, meaning *Wooden Town* in Serbian, is a traditional village that the Serbian film director Emir Kusturica had built for his film *Life Is a Miracle*. It is located in the Zlatibor District near the city of Užice, two hundred kilometers southwest of Serbia's capital, Belgrade. It is located near Mokra Gora and Višegrad.

Slovakia

- Partizánske was established in 1938–1939, when Jan Antonín Baťa of Zlín, Czechoslovakia (now Czech Republic) and his powerful network of companies built a shoe factory in the cadastral area of Šimonovany municipality. The newly created settlement for workers carried the name of Baťovany and was part of

Šimonovany. With the growth of the factory, so grew the settlement. The whole municipality was renamed to Baťovany in 1948 and given town status. As a sign of recognition of local inhabitants fighting in the Slovak National Uprising, the town was renamed Partizánske on 9 February 1949.

- Svit was established in 1934 by business industrialist Jan Antonín Baťa of Zlín, Czechoslovakia (now Czech Republic) in accordance with his policy of setting up villages around the country for his workers.

Slovenia

Nova Gorica, built after 1947 immediately to the east of the new border with Italy, in which the town of Gorizia remained.

Spain

During the 16th and 17th centuries, the population of Spain declined due to emigration to the Americas and later kings and governments made efforts to repopulate the country. In the second half of the 18th century, King Charles III implemented the so-called New Settlements (Nuevas Poblaciones) plan which would bring 10,000 immigrants from central Europe to the region of Sierra Morena. Pablo de Olavide was appointed superintendent and about forty new settlements were established of which the most notable was La Carolina, which has a perfectly rectangular grid design.

Later kings and repopulation efforts led to the creation of more settlements, also with rectangular grid plans. One of them was the town of La Isabela (40.4295 N, 2.6876 W), which disappeared in the 1950s submerged under the waters of the newly created artificial lake of Buendía but is still visible just under the water in satellite imagery.

Under Francisco Franco, the Instituto Nacional de Colonización (National Institute of Colonization) built a great number of towns and villages.

Tres Cantos, near Madrid, is a good example of a successful new town design in Spain. It was built in the 1970s.

Newer additional sections of large cities are often newly planned as is the case of the Salamanca district or Ciudad Lineal in Madrid or the Eixample in Barcelona.

Sweden

Gothenburg was planned and built as a major fortified city from nothing from 1621.

Karlskrona was also planned and built as a major city and naval base from nothing, beginning 1680.

Vällingby, a suburb, is an example of a new town in Sweden from after 1950.

Kiruna was built because of the large mine, from 1898.

Arvika was also a planned city.

Turkey

Some parts of the biggest city, Istanbul, being re-developed and re-planned.

The capital, Ankara, was built by a plan and constantly re-planned.

Atça, Aydın was burned down by Greek forces in Turkish Independence War. Rebuilding plan was based on Paris' plan.

Some other cities including; Erzincan, Karabük, Kars, Kayseri, Konya were also planned.

Ukraine

Odesa was built as a planned city according to 18th-century plans by the Flemish engineer Franz de Wollant (also known as François Sainte de Wollant). The same engineer also planned the following municipalities in Ukraine in the late 18th century:

- Voznesensk (Ukrainian: Вознесенськ), in Mykolayiv Oblast
- Ovidiopol (Ukrainian: Овідіополь), in Odesa Oblast

Komsomolsk, founded in the 1960, is the most prosperous planned city in Ukraine, depending on the internationally-important iron ore mining business.

Prypiat is another new city in Ukraine built in 1970. The city was abandoned on April 27, 1986 after the Chernobyl nuclear disaster. On April 26 the city had 50,000 habitants, the majority working at the Chernobyl nuclear power plant. Now the abandoned town is highly contaminated by radiation. Most of the Prypiat's former inhabitants were resettled to Slavutych which was planned and built for that purpose.

United Kingdom

England and Scotland

The Romans planned many towns in Britain, but the settlements were changed out of all recognition in subsequent centuries. The town of Winchelsea is said to be the first post-Roman new town in Britain, constructed to a grid system under the instructions of King Edward I in 1280, and largely completed by 1292. Another claimant to the title is Salisbury, established in the early 13th century by the then Bishop of Sarum. The best known pre-20th-century new town in the UK was undoubtedly the Edinburgh New Town, built in accordance with a 1766 master plan by James Craig, and (along with Bath and Dublin) the archetype of the elegant Georgian style of British architecture.

The term "new town" often refers in the UK to towns built after World War II under the New Towns Act 1946. These were influenced by the garden city movement, launched around 1900 by Ebenezer Howard and Sir Patrick Geddes and the work of Raymond Unwin, and manifested at Letchworth Garden City and Welwyn Garden City in Hertfordshire.

Following World War II, some 28 projected towns were designated as New Towns under the 1946 Act, and were developed partly to house the large numbers of people whose homes had been destroyed by bombing during the war and partly to move parts of the population out of (mainly Victorian) urban slums. New Towns policy was also informed by a series of wartime commissions, including:

- the Barlow Commission (1940) into the distribution of industrial population,
- the Scott Committee into rural land use (1941)
- the Uthwatt Committee into compensation and betterment (1942)
- (later) the Reith Report into New Towns (1947).

Also crucial to thinking was the Abercrombie Plan for London (1944), which envisaged moving a million and a half people from London to new and expanded towns. A similar plan was developed for the Clyde Valley in 1946 to combat similar problems faced in Glasgow. Together these committees reflected a strong consensus to halt the uncontrolled sprawl of London and other large cities. For some, this consensus was tied up with a concern for social welfare reform (typified by the Beveridge Report), as typified in the motto *if we can build better, we can live better*; for others, such as John Betjeman it was a more conservative objection to the changing character of existing towns.

Following the building of Borehamwood, Middlesex, 12 miles north-west of central London, the first in a ring of major "first generation" New Towns around London (1946) were Stevenage, Hertfordshire, 27 miles to the north of London, and Basildon, Essex, 25 miles east of London along the River Thames. Hertfordshire built four other new towns, two in the vicinity of Stevenage (Welwyn Garden City and Hatfield), a third to the north called Letchworth, and Hemel Hempstead to the west. (Hall 1996: 133) New Towns in the North East were also planned such as Newton Aycliffe (which the social reformer and government adviser William Beveridge wanted to be the "ideal town to live in") and Peterlee. Two new towns were also planned in Scotland at East Kilbride (1947) and Glenrothes (1948). Bracknell in Berkshire, to the south-west of London, was designated a New Town in 1949 and is still expanding. Other London new towns from this era include Harlow in Essex and Crawley in West Sussex.

Later a scatter of "second-generation" towns were built to meet specific problems, such as the development of the Corby Steelworks. Finally, five "third-generation" towns were launched in the late 1960s: these were larger, some of them based on substantial existing settlements such as Peterborough, and the most famous was probably Milton

Keynes, midway between London and Birmingham, known for its huge central park and shopping centre, designed from the outset as a new city – though in law it is a 'New Town'. The 1960s saw new towns crop up around England's second-city Birmingham, namely Redditch, Tamworth and Telford.

Other towns, such as Ashford in Kent, Basingstoke in Hampshire and Swindon in Wiltshire, were designated "Expanded Towns" and share many characteristics with the new towns. Scotland also gained three more new towns: Cumbernauld in 1956, famous for its enclosed 'town centre', Livingston (1962) and Irvine (1966).

In spite of the relative success of new towns in the London Metropolitan green belt, London continued to suffer from a chronic housing shortage, especially in the south-east. Another small New Town, Thamesmead, was developed adjacent to the Thames in the early 1960s but suffered from poor transport links. Some improvement in infrastructure has been seen subsequently.

All the new towns featured a car-oriented layout with many roundabouts and a grid-based road system unusual in the old world. Milton Keynes in particular was designed with a grid-based distributor road system. The earlier new towns, where construction was often rushed and whose inhabitants were generally plucked out of their established communities with little ceremony, rapidly got a poor press reputation as the home of "new town blues". These issues were systematically addressed in the later towns, with the third generation towns in particular devoting substantial resources to cycle routes, public transport and community facilities, as well as employing teams of officers for social development work.

The financing of the UK new towns was creative. Land within the designated area was acquired at agricultural use value by the development corporation for each town, and infrastructure and building funds borrowed on 60-year terms from the UK Treasury. Interest on these loans was rolled up, in the expectation that the growth in land values caused by the development of the town would eventually allow the loans to be repaid in full. However, the high levels of retail price inflation experienced in the developed world in the 1970s and 1980s fed through into interest rates and frustrated this expectation, so that substantial parts of the loans had ultimately to be written off.

All New Towns designated under the New Towns Act of 1946 were serviced by a secretariat, the New Towns Association, a quango that reported to the New Towns Directorate of the Department of the Environment. It coordinated the work of the General Managers and technical officers, published a monthly information bulletin and provided information for visitors from around the world. As each New Town reached maturity, the town's assets were taken over by the Commission for New Towns. Set up in 1948, the New Towns Association was dissolved in 1998. All papers held by it and the Commission for New Towns are held in The National archives:

From the 1970s the first generation towns began to reach their initial growth targets. As they did so, their development corporations were wound up and the assets disposed of: rented housing to the local authority, and other assets to the Commission for the New Towns (in England; but alternative arrangements were made in Scotland and Wales). The Thatcher Government, from 1979, saw the new towns as a socialist experiment to be discontinued, and all the development corporations were dissolved by 1992 (with the closure of Milton Keynes Development Corporation), even for the third generation towns whose growth targets were still far from being achieved. Ultimately the Commission for the New Towns was also dissolved and its assets - still including a lot of undeveloped land - passed to the English Industrial Estates Corporation (later known as English Partnerships).

Many of the New Towns attempted to incorporate public art and cultural programmes but with mixed methods and results. In Harlow the architect in charge of the design of the new town, Frederick Gibberd, founded the Harlow Art Trust and used it to purchase works by leading sculptors, including Auguste Rodin, Henry Moore and Barbara Hepworth. In Peterlee the abstract artist Victor Pasmore was appointed part of the design team, which led to the building of the Apollo Pavilion. Washington New Town was provided with a community theatre and art gallery. The public art in Milton Keynes includes the (in)famous Concrete Cows, which resulted from the work of an 'artist in residence' and have gone on to become a recognised landmark. Glenrothes led the way in Scotland being the first new town to appoint a town artist in 1968. A massive range of artworks (around 132 in total) ranging from concrete hippos to bronze statues, dancing children, giant flowers, a dinosaur, a horse and chariot and crocodiles, to name but a few, were created. Town artists appointed in Glenrothes include David Harding and Malcolm Robertson.

In the 1990s, an experimental "new town" was developed by the Prince of Wales to use very traditional or vernacular architectural styles was started at Poundbury in Dorset.

Northern Ireland

In Northern Ireland, building of Craigavon in County Armagh commenced in 1966 between Lurgan and Portadown, although entire blocks of flats and shops lay empty, and later derelict, before eventually being bulldozed. The area, which now has a population exceeding 80,000 is mostly a dormitory town for Belfast.

Derry was the first ever planned city in Ireland (the city is now located in Northern Ireland). Work began on building the new city across the River Foyle from the ancient town of Derry (*Doire Cholm Chille* or *Doire*) in 1613. The walls were actually completed five years later in 1618. The central diamond within a walled city with four gates was thought to be a good design for defence. In 1963 under the Matthew Plan the new city of Craigavon was founded out of the original towns of Portadown and Lurgan. This town today lies mostly incomplete as the troubles halted construction. The plan initially was to construct a relief settlement to take people out of the crowded city of Belfast.

Scotland

In the late 1950s and early 1960s Scotland saw a creation of several "post-war new towns". These were; Cumbernauld, East Kilbride, Glenrothes, Irvine and Livingston. Each of these towns are in Scotland's list of 20 most populated towns and cities. East Kilbride is the second largest town in Scotland, or the 6th largest settlement with a population of over 73,000 and Livingston with a population of 76,000. The other three towns are not as big with populations between 30,000 and 50,000.

Wales

The only new towns in Wales have been Newtown and Cwmbran. Cwmbran was established to provide new employment in the south eastern portion of the South Wales Coalfield. The town is perhaps most widely known now for its international sports stadium and shopping centre.

North America

Canada

When Prime Minister Sir John A. Macdonald began to settle the West in Canada, he put the project under the command of the Canadian Pacific Railway (CPR). The CPR exercised complete control over the development of land under its ownership. The federal government granted every second square mile section (totalling 101,000 km^2) along the proposed railway line route to the CPR. The CPR decided where to place railway stations, and thus would decide where the dominant town of the area would be. In most instances the CPR would build a station on an empty section of land to make the largest profit from land sales – meaning that the CPR founded many of the Canadian West's towns, such as Medicine Hat and Moose Jaw, from scratch. If an existing town was close to the newly constructed station but on land not owned by the CPR, the town was forced to move itself to the new site and reconstruct itself, essentially building a new town. Calgary and Yorkton, Saskatchewan, were among the towns that had to move themselves.

After the CPR established a station at a particular site, it would plan how the town would be constructed. The side of the tracks with the station would go to business, while the other side would go to warehouses. Furthermore, the CPR controlled where major buildings went (by giving the town free land to build it where the CPR wanted it to go), the construction of roads and the placement and organization of class-structured residential areas.

The CPR's influence over the development of the Canadian west's communities was one of the earliest examples of new town construction in the modern world. Later influences on planned community development in Canada were the exploitation of mineral and forest wealth, usually in remote locations of the vast country. Among numerous company towns planned and built for these purposes were Corner Brook and Grand Falls in Newfoundland, Témiscaming and Fermont in Quebec.

In the modern suburban context, several "New Towns" were established in the suburbs of large cities. Early examples include Leaside in Toronto and Mount Royal in Montreal. Both were planned and developed by the Canadian Northern Railway as middle class suburbs, though both, Leaside in particular, featured large industrial tracts. Leaside had its own municipal government until 1967, while Mount Royal continues to enjoy autonomy from the City of Montreal.

In the post-war period, new corporate new towns were developed. Bramalea, located in Brampton, Ontario and Erin Mills, located in Mississauga, Ontario, were both developed in phases. Both included residential, commercial and industrial components. Development in Erin Mills continues to this day.

More recently, the Cornell development in Markham, Ontario, was built as a new town, using the concepts of New Urbanism.

Mexico

Tenochtitlan was the capital of the Aztec empire, which was built on an island in Lake Texcoco in what is now the Federal District in central Mexico. The city was largely destroyed in the 1520s by Spanish conquistadores. Mexico City was erected on top of the ruins and, over the ensuing centuries, most of Lake Texcoco has gradually been drained.

Puebla was built because of the need of a Spanish settlement in the route between Mexico City and the port of Veracruz.

Panama

Although Panama City itself is not planned, certain areas are such as Costa del Este, an exclusive high density residential and business area, very close to downtown Panama City. The project combines many skyscrapers with beautiful green areas, and it is close to a highway that connects it to the city center. Other planned areas, but in a lesser degree, are Punta Pacifica and the former Canal Zone.

United States

In the colonial history of the United States, the first planned community was St. Augustine, planned in 1565. The earliest towns in English-speaking America such as Jamestown had only rudimentary elements of planning. The first comprehensively-planned town was Charles Town (later Charleston, South Carolina), which was founded in 1670, planned in 1672, and relocated in 1680. Later planned cities were Philadelphia, 1682; Albany, 1695; Williamsburg, 1699; Annapolis, 1718; New York 1731 (redesigned by the British);and Savannah, 1733; New Haven, 1748 (with an early plan dated 1638); and Alexandria, 1749. The national capital (Washington, D.C.), and several state capitals (Jackson, Mississippi; Columbus, Ohio; Indianapolis, Indiana; Raleigh, North Carolina; Columbia, South Carolina; Madison, Wisconsin; Salt Lake City, Utah; Tallahassee, Florida; and Austin, Texas) were essentially carved out of the wilderness to serve as capital cities.

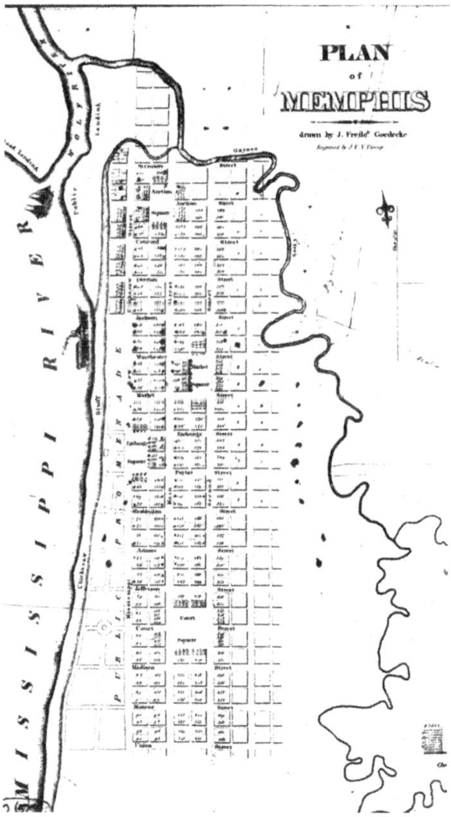

The original plan for Memphis, Tennessee, as surveyed in 1819

Pullman, now incorporated into Chicago's South Side, was a world-renowned company town founded by the industrialist George M. Pullman in the 1880s. In Beaver County, Pennsylvania near Pittsburgh, American Bridge Company founded Ambridge, Pennsylvania in 1905 as a company town for American Bridge; American Bridge is still based near Ambridge today in nearby Coraopolis, Pennsylvania.

Another well known company town is Gary, Indiana, which was founded in 1906 by the United States Steel Corporation as a home for its new steel mill, the Gary Works, and named after Elbert Henry Gary, the chairman of the company. For many years the Gary Works was the largest steel mill in the world, and it dominated the town, the main entrance being at the northern end of Broadway, the city's main thoroughfare. The fortunes of this planned city have historically risen and fallen with those of the steel mill: prosperous in the 1930s, the city has lost 55 percent of its population since 1960.

Riverside, Illinois, Radburn, New Jersey, and Kansas City, Missouri's Country Club District are other early examples of planned communities. Established in 1912, Shaker Heights, Ohio, was planned and developed in by the Van Sweringen brothers, railroad moguls who envisioned the community as a suburban retreat from the industrial inner-city of Cleveland. Kohler Company created a planned village of the same name west of the company's former headquarters city of Sheboygan, Wisconsin, which incorporated in 1912. In 1918, the Aluminum Company of America built the town of Alcoa, Tennessee for the employees of the nearby aluminum processing plant.

During the Florida land boom of the 1920s in Southern Florida, the communities of Coral Gables, Opa-locka, and Miami Springs, now suburbs of Miami, were incorporated as fully planned "themed" communities which were to reflect the architecture and look of Spain, Arabia, and Mexico respectively, and are now considered some of the first modern planned communities in the United States. Oldsmar, located in west central Florida, was developed by automobile pioneer Ransom E. Olds.

In 1928, San Clemente, California was incorporated by Ole Hanson who designated that all buildings must be approved by an architectural review board in order to retain control over development and building style.

During the Great Depression of the 1930s, several model towns were planned and built by the Federal government. Arthurdale and Eleanor, West Virginia, federally funded New Deal communities, were Eleanor Roosevelt's projects to ease the burden of the depression on coal miners. The Tennessee Valley Authority created several towns of its own to accommodate workers constructing their new dams; the most prominent being Norris, Tennessee. Three "Greenbelt Communities", Greenbelt, Maryland, Greenhills, Ohio, and Greendale, Wisconsin, built by the Federal government during the 1930s were planned with a surrounding "belt" of woodland and natural landscaping.

During World War II, the Manhattan Project built several planned communities to provide accommodations for scientists, engineers, industrial workers and their families. These communities, including Oak Ridge, Tennessee, Richland, Washington and Los Alamos, New Mexico were necessary because the laboratories and industrial plants of the Manhattan Project were built in isolated locations to ensure secrecy. Even the existence of these towns was a military secret, and the towns themselves were closed to the public until after the war.

The Levittowns—in Long Island, Pennsylvania and New Jersey (now known as Willingboro, New Jersey) -- typified the planned suburban communities of the 1950s and early 1960s. California's Rohnert Park (north of San Francisco) is another example of a planned city (built at the same time as Levittown) that was marketed to attract middle-class people into an area only populated with farmers with the phrase, "A Country Club for the middle class."

Aerial view of Levittown, Pennsylvania circa 1959

Many other places, such as Orange County, California, the Conejo Valley in Ventura County, Valencia in Los Angeles County, as well as Phoenix, Arizona and Northern Arizona also have many master planned communities following the housing boom in the 1960s, which is when the fathers of Scottsdale, Arizona foresaw a huge amount of growth in Arizona. Some of those communities include Anaheim Hills, Rossmoor, Irvine, Ladera Ranch, Laguna Niguel, Mission Viejo, and Talega, Thousand Oaks, Westlake Village, Newbury Park, Valencia in California and (in the Phoenix area) Marley Park, Talking Rock Ranch, McCormick Ranch, Rio Verde, Tartesso and Verrado in Buckeye, Arizona. The neighborhood of Warren in the city of Bisbee has the distinction of being Arizona's first planned community. In the Conejo Valley, which is the in East County Area of Ventura County, all cities were master planned. Most notably, the Thousand Oaks, Newbury Park, and Westlake Village area was master planned by the Janss Investment Company, which was also responsible for the development of Westwood Village, part of the Westside in Los Angeles. Valencia is an area that is a master planned community that incorporated into the City of Santa Clarita, developed and planned by the Newhall Land and Farming Company. About 25% of Orange County is composed of various master planned communities, much of which was done by the Irvine Company, and since 1990, 85% of all developments in Orange County and a slightly smaller amount of communities in Arizona were part of a master planned community. 75% of all resales today in the Phoenix area are homes in master planned communities, and 80% of all new home construction permits issued by Arizona building departments are master planned communities. These communities provide functionality to the precious land left in the area, as well as the ability to create a housing-business-transportation-open space balance.

The era of the modern planned city began in 1962-64 with the creation of Reston, Virginia which was begun just a year before Coral Springs in western Broward County, Florida, and Columbia, Maryland. In more recent years, New Urbanism has set the stage for new cities, with places like the idyllic Seaside, Florida, and Disney's new town of Celebration, Florida.

In the United States, suburban growth in the Sunbelt states has coincided with the popularity of Master Planned Communities within established suburbs. Texas was at the forefront of this trend. Las Colinas, established in 1973, was one of the first such examples and is still growing. Las Colinas is a 12,000 acres (4,900 ha) master planned community within the Dallas-area city of Irving. In 2006, residents approved changes to deed restrictions to allow greater density of urban mixed-use and residential construction. Also in the 1970s, just north of the existing town of Spring, Texas (north of Houston), oil and gas industry executive George P. Mitchell developed The Woodlands, a major residential and commercial master planned community which is now considered one of the premier residential and business destinations in the Houston area. The Woodlands is still experiencing huge growth to this day. In the 1990s, Cinco Ranch was first developed just south of the existing town of Katy, one of the western suburbs of Houston, and has contributed to the explosive recent growth of Houston's suburban west side.

In the San Francisco Bay Area, master planned commercial developments such as Bishop Ranch in San Ramon and Hacienda Business Park in Pleasanton have attracted major corporate tenants to relocate from downtown Oakland and San Francisco; these companies include Safeway, Chevron Corporation and AT&T (as the former Pacific Bell).

In recent years, new towns such as Mountain House, San Joaquin County, California, have added a new wrinkle to the movement: to prevent conurbation with nearby cities, they have imposed strict growth boundaries, as well as automatic "circuit breakers" that place moratoriums on residential development if the number of jobs per resident in the town falls below a certain value. Centennial new town part in Tejon Ranch halfway between Los Angeles and Bakersfield, will incorporate such restrictions in order to minimize the commuter load on severely congested I-5. Coyote Springs, Nevada, Destiny, Florida and Douglas Ranch in Buckeye, Arizona are amongst the largest communities being planned for the 21st century. A recent twist is the town of Ave Maria, Florida, founded in 2007, which is anchored by a Catholic university and has a large Catholic church in the center of town surrounded by commercial and residential development.

South America

Argentina

La Plata was planned in 1880 to replace Buenos Aires city as the capital of the Buenos Aires Province.

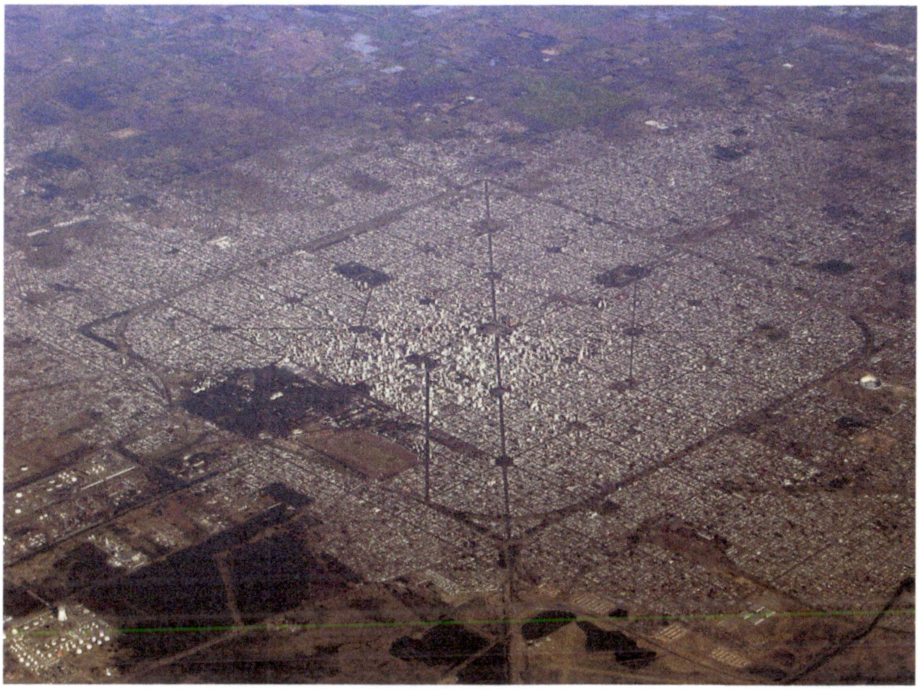

La Plata from the air.

Urban planner Pedro Benoit designed a city layout based on a rationalist conception of urban centers. The city has the shape of a square with a central park and two main diagonal avenues, north-south and east-west. (In addition, there are numerous other shorter diagonals.) This design is copied in a self-similar manner in small blocks of six by six blocks in length. Every six blocks, one finds a small park or square. Other than the diagonals, all streets are on a rectangular grid, and are numbered consecutively.

The designs for the government buildings were chosen in an international architectural competition. Thus, the Governor Palace was designed by Italians, City Hall by Germans, etc. Electric street lighting was installed in 1884, and was the first of its kind in Latin America.

Brazil

Brasília:

Juscelino Kubitschek, President of Brazil from 1956 to 1961, ordered the construction of Brasília, fulfilling the promise of the Constitution and his own political campaign promise. Building Brasília was part of Juscelino's "fifty years of prosperity in five" plan. Lúcio Costa won a contest and was the main urban planner in 1957, with 5550 people competing. Oscar Niemeyer, a close friend, was the chief architect of most public buildings and Roberto Burle Marx was the landscape designer. Brasília was built in 41 months, from 1956 to April 21, 1960, when it was officially inaugurated.

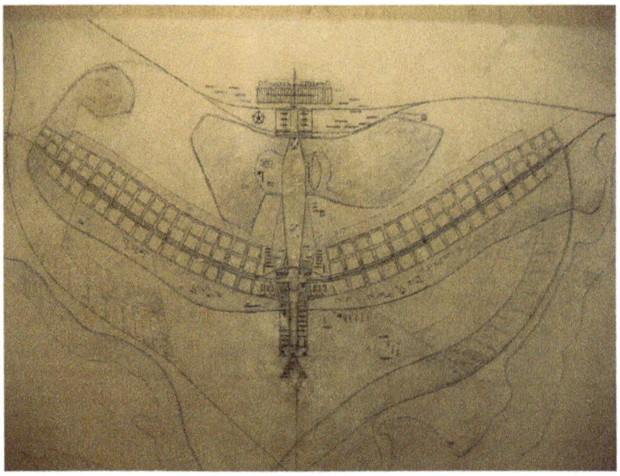

Brasília: Pilot Plan

The former capital of Brazil was Rio de Janeiro, and resources tended to be concentrated in the southeast region of Brazil. While the city was built because there was a need for a neutrally-located federal capital, the main reason was to promote the development of Brazil's hinterland and better integrate the entire territory of Brazil. Brasília is approximately at the geographical center of Brazilian territory.

Lúcio Costa, the city's principal architect, designed the city to be shaped like an airplane. Housing and offices are situated on giant superblocks, everything following the original plan. The plan specifies which zones are residential, which zones are commercial, where industries can settle, where official buildings can be built, the maximum height of buildings, etc.

Belo Horizonte:

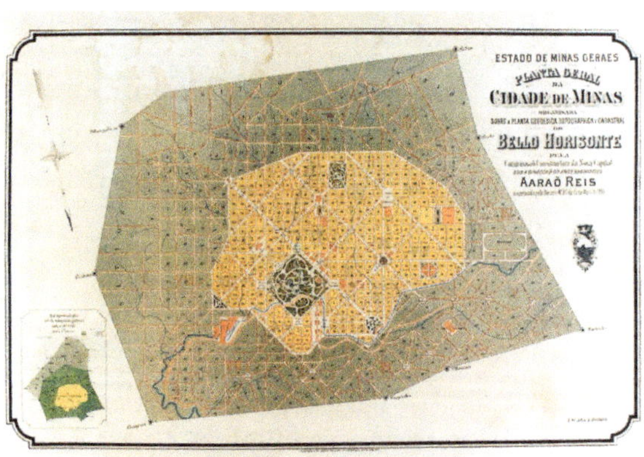

Belo Horizonte in 1895.

In 1889, Brazil became a republic, and it was agreed that a new state capital of Minas Gerais, in tune with a modern and prosperous Minas Gerais, had to be set. In 1893, due to the climatic and topographic conditions, Curral Del Rey was selected by Minas Gerais governor Afonso Pena among other cities as the location for the new economical and cultural center of the state, under the new name of "Cidade de Minas," or City of Minas. Aarão Reis, an urbanist from the State of Pará, was then set to design the second planned city of Brazil (the first one is Teresina), and then Cidade de Minas was inaugurated finally in 1897, with many unfinished constructions as the Brazilian Government set a deadline for its completion. Inhabitation of the city was subsidized by the local government, through the concession of free empty lots and funding for building houses. An interesting feature of Reis' downtown street plan for Belo Horizonte was the inclusion of a symmetrical array of perpendicular and diagonal streets named after Brazilian states and Brazilian indigenous tribes.

Goiânia: The plan was for a city of 50,000 with the shape of a concentric radius – streets in the form of a spoke, with the Praça Cívica as the center, with the seats of the state and municipal government – The Palace of Emeralds and the Palace of Campinas. In 1937, a decree was signed transferring the state capital from the Cidade de Goiás to Goiânia. The official inauguration only occurred in 1942 with the presence of the president of the republic, governors, and ministers.

Fordlândia was built to be a part of Henry Ford's motor company. Originally intended to be a rubber plantation, it failed within several years and is now home to squatting farmers.

Other notable planned cities in Brazil include Teresina (The first one, inaugurated in 1842), Petrópolis, Boa Vista, Palmas, Londrina, and Maringá (the latter two in the state of Paraná).

Venezuela

- Guayana City - Bolívar State
- Puerto Ordaz - Bolívar State

Oceania

Australia

Australia's most prominent fully planned city is Canberra designed by American architect Walter Burley Griffin. The early central areas of two other major capital cities — Adelaide and Melbourne — were also planned by surveyors. Walter Burley Griffin was Australia's most notable city planner having also designed smaller cities and towns including Leeton and Griffith in New South Wales. A controversial Japanese-backed planned city, Multifunction Polis, was proposed in the 1980s but never implemented.

Australia is still building planned communities with developers such as Delfin Lend Lease, an S&P/ASX 50 company, that has been responsible for large master planned communities such as;

- Forest Lake, Queensland, Brisbane (completed 2004)
- The New Rouse Hill, Sydney (current)
- Golden Grove, South Australia (Completed 1991)

Adelaide

Adelaide's planned town grids were surrounded by parkland and intersected by the River Torrens

Adelaide was founded by British and German colonists in 1836 to test out Edward Gibbon Wakefield's theories of systematic colonisation. Convict labour was not employed and the colony in theory would be financially self-sufficient; in practice,government assistance was used in the early stages. Land had been sold before anyone set foot in the largely unexplored territory and the city (the basis for the future central business district) was surveyed and planned in a remarkably short space of time. Adelaide's design has been praised for its four-square layout, its choice of setting and its ample parklands which have had minimal encroachment of developments. The town centre was in sufficient proximity to a water source, the River Torrens.

Melbourne

A reconstruction of Robert Hoddle's original plans for Melbourne's central grid which defined the early township and today's city centre

Melbourne was planned as a free settlement in 1837 through the Hoddle Grid, drawn up by Robert Hoddle under instructions from George Gipps, the original plan for Melbourne as part of the first land sales (prior to the planning only a handful of existing settlements were built on the fringe of the grid). The grid featured wide parallel streets, spanning a gently sloping valley between hills (Batman's Hill, Flagstaff Hill and Eastern Hill) and roughly parallel to the course of the Yarra River. The deliberate exclusion of city squares or open space within the grid was a subject of future frustration for the municipality and residents Elizabeth Street, Melbourne, in the centre of the grid, was built over a gully and has therefore been prone to flooding. Despite a later extension and later inclusion of planned suburbs, Melbourne's original plans were not as extensive as Adelaide's, and the city rapidly outgrew its original boundaries. As such, it is often not considered to be a planned city, but the grid continues to define much of the character of the Melbourne city centre.

Canberra

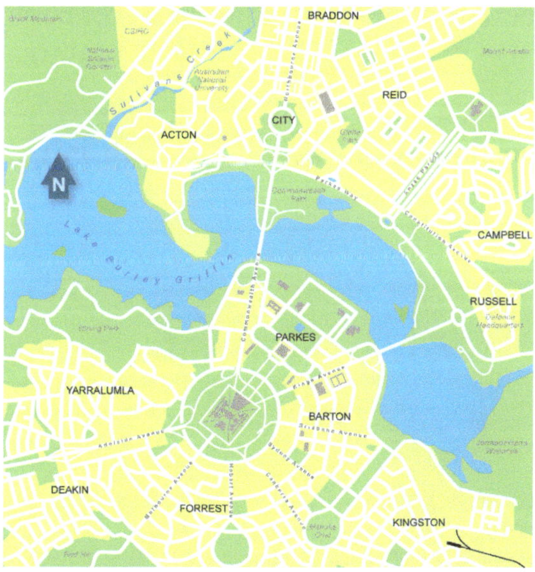

Inner Canberra demonstrates some aspects of the Griffin plan, in particular the Parliamentary Triangle

Canberra, established in 1908, was planned as the capital city of Australia and is Australia's most notable and ambitious example of civic planning. The city was designed to be the Federal Capital following the federation of the six Australian colonies which formed the Commonwealth of Australia. The new nation required a capital that was located away from other major settlements such as Melbourne and Sydney. Canberra is thus located in a Territory - the Australian Capital Territory - and not a State. Prior to this time the land that Canberra is found on was farming land, indigenous settlements, and forest. In 1912, after an extensive planning competition was completed, the vision of American Walter Burley Griffin was chosen as the winning design for the city. Extensive construction and public works were required to complete the city, this involved the flooding of a large parcel of land to form the center piece of the city, Lake Burley Griffin. Unlike some other Australian cities, the road network, suburbs, parks and other elements of the city were designed in context with each other, rather than haphazard planning as witnessed in much of Sydney. Notable buildings include the High Court, Federal Parliament, Government House, War Memorial, Anzac Parade and headquarters of the Department of Defence.

New Zealand

New Zealand has several small New Towns, built for a specific purpose. Examples include Kawerau in the Bay of Plenty (a mill town), Twizel in North Otago, Mangakino in the Waikato (both for hydroelectricity), and Turangi near Taupo (for the Tongariro Power Scheme). Construction of Kawerau began in 1953. Twizel was built in 1968 to service nearby hydroelectric projects and was supposed to close on their completion. However, its residents managed to save the town in 1983. Mangakino, constructed from 1946, was also meant to be a temporary construction town, but it too remains today. John Martin, the founder of the Wairarapa town of Martinborough, set out the town's first streets in the pattern of the Union Flag in the 19th century.

In 2006, construction began on Pegasus Town, a new planned town adjacent to Woodend, New Zealand approximately 25 km (16 mi) north of Christchurch.

Park

Veale Gardens in Australia.

A park is an area of natural, semi-natural, or planted space set aside for human enjoyment and recreation or for the protection of wildlife or natural habitats. It may consist of grassy areas, rocks, soil, and trees, but may also contain buildings and other artifacts such as monuments, fountains or playground structures. In North

America, many parks have fields for playing sports such as soccer, baseball and football, and paved areas for games such as basketball. Many parks have trails for walking, biking and other activities. Some parks are built adjacent to bodies of water or watercourses, and these parks may comprise a beach or boat dock area. Often, the smallest parks are in urban areas, where a park may take up only a city block or less. Urban parks often have benches for sitting and they may contain picnic tables and barbecue grills. Parks have differing rules regarding whether dogs can be brought into the park: some parks prohibit dogs; some parks allow them with restrictions (e.g., use of a leash); and some parks, which may be called "dog parks," permit dogs to run off-leash.

The largest parks can be vast natural areas of hundreds of thousands of square kilometres (thousands of square miles), with abundant wildlife and natural features such as mountains and rivers. In many large parks, camping in tents is allowed with a permit. Many natural parks are protected by law, and users may have to follow restrictions (e.g., rules against open fires or bringing in glass bottles). Large national and sub-national parks are typically overseen by a park ranger or a park warden. Large parks may have areas for canoeing and hiking in the warmer months and, in some northern hemisphere countries, cross-country skiing and snowshoeing in colder months.

History

Depiction of a medieval hunting park from a 15th-century manuscript

The first parks were English deer parks, land set aside for hunting by royalty and the aristocracy in medieval times. They had walls or thick hedges around them to keep game animals (e.g., stags) in and people out. It was strictly forbidden for commoners to hunt animals in these deer parks.

These game preserves evolved into landscaped parks set around mansions and country houses from the sixteenth century onwards. These may have served as hunting grounds but they also proclaimed the owner's wealth and status. An aesthetic of landscape design began in these stately home parks where the natural landscape was enhanced by landscape architects such as Capability Brown. As cities became crowded, the private hunting grounds became places for the public.

With the Industrial revolution parks took on a new meaning as areas set aside to preserve a sense of nature in the cities and towns. Sporting activity came to be a major use for these urban parks. Areas of outstanding natural beauty were also set aside as national parks to prevent their being spoiled by uncontrolled development.

In some parks or time periods with high pollen counts, parks tend to be avoided.

Design

Park design is influenced by the intended purpose and audience, as well as by the available land features. A park intended to provide recreation for children may include a playground. A park primarily intended for adults may feature walking paths and decorative landscaping. Specific features, such as riding trails, may be included to support specific activities.

The design of a park may determine who is willing to use it. Walkers may feel unsafe on a mixed-use path that is dominated by fast-moving cyclists or horses. Different landscaping and infrastructure may even affect children's rates of use of parks according to gender. Redesigns of two parks in Vienna suggested that the creation of multiple semi-enclosed play areas in a park could encourage equal use by boys and girls.

Parks are part of the urban infrastructure: for physical activity, for families and communities to gather and socialize, or for a simple respite. Research reveals that people who exercise outdoors in green-space derive greater mental health benefits. Providing activities for all ages, abilities and income levels is important for the physical and mental well-being of the public.

Design for Safety

Parks need to feel safe for people to use them. Research shows that perception of safety can be more significant in influencing human behaviour than actual crime statistics. If citizens perceive a park as unsafe, they might not make use of it at all.

A well-lit path in Dehli's Garden of Five Senses

There are a number of features that contribute to whether or not a park feels safe. Elements in the physical design of a park, such as an open and welcoming entry, good visibility (sight lines), and appropriate lighting and signage can all make a difference. Regular park maintenance, as well as programming and community involvement can also contribute to a feeling of safety.

Women as a Measure of Safety

In the United States, the standard for safety in parks is increasingly measured by whether women feel safe in that particular location. This was originally identified by the urban sociologist William H Whyte in his studies decades ago in New York. Research reveals that women have a different sense of safety compared to men, whether they are walking in their neighborhood or in a park. Dan Biederman, President of the Bryant Park Corp. stated "Women pick up on visual cues of disorder better than men do.... And if women don't see other women, they tend to leave." Whether or not a woman feels safe can determine how much physical activity she has and if it will reach the recommended level for good health and disease prevention. Park designers and planners can take several steps to increase women's safety from sexual assault or other assault, including providing sufficient lighting, having patrols by police officers or other safety officials, and providing emergency buttons for summoning assistance.

Active and Passive Recreation Areas

Parks can be divided into active and passive recreation areas. Active recreation is that which has an urban character and requires intensive development. It often involves cooperative or team activity, including playgrounds, ball fields, swimming pools, gymnasiums, and skateparks. Active recreation such as team sports, due to the need to provide substantial space to congregate, typically involves intensive management, maintenance, and high costs. Passive recreation, also called "low intensity recreation" is that which emphasizes the open-space aspect of a park and allows for the preservation of natural habitat. It usually involves a low level of development, such as rustic picnic areas, benches and trails.

Many smaller *neighborhood parks* are receiving increased attention and valuation as significant community assets and places of refuge in heavily populated urban areas. Neighborhood groups around the world are joining together to support local parks that have suffered from urban decay and government neglect.

Passive recreation typically requires little management and can be provided at very low costs. Some open space managers provide nothing other than trails for physical activity in the form of walking, running, horse riding, mountain biking, snow shoeing, or cross-country skiing; or sedentary activity such as observing nature, bird watching, painting, photography, or picnicking. Limiting park or open space use to passive recreation over all or a portion of the park's area eliminates or reduces the burden of managing active recreation facilities and developed infrastructure. Many ski resorts combine active recreation facilities (ski lifts, gondolas, terrain parks, downhill runs, and lodges) with passive recreation facilities (cross-country ski trails).

Parks Owned or Operated by Government

National Parks

Northeast Greenland National Park, the world's largest national park

A national park is a reserve of land, usually, but not always declared and owned by a national government, protected from most human development and pollution. Although this may be so, it is not likely that the government of a specific area owns it, rather the community itself. National parks are a protected area of International Union for Conservation of Nature Category II. This implies that they are wilderness areas, but unlike pure nature reserves, they are established with the expectation of a certain degree of human visitation and supporting infrastructure.

While this type of national park had been proposed previously, the United States established the first "public park or pleasuring-ground for the benefit and enjoyment of the people", Yellowstone National Park, in 1872, although Yellowstone was not gazetted as a national park. The first officially designated national park was Mackinac Island, gazetted in 1875. Australia's Royal National Park, established in 1879, was the world's second officially established national park.

The largest national park in the world is the Northeast Greenland National Park, which was established in 1974 and currently protects 972,001 km² (375,000 sq mi)

Sub-National Parks

In some Federal systems, many parks are managed by the sub-national levels of government. In Brazil, the United States, and some states in Mexico, as well as in the Australian state of Victoria, these are known as state parks, whereas in Argentina, Canada and South Korea, they are known as provincial or territorial parks.

Urban Parks

Yoyogi Park is a large urban park in Tokyo.

A park is an area of open space provided for recreational use, usually owned and maintained by a local government. Parks commonly resemble savannas or open woodlands, the types of landscape that human beings find most relaxing. Grass is typically kept short to discourage insect pests and to allow for the enjoyment of picnics and sporting activities. Trees are chosen for their beauty and to provide shade.

Sad Janka Kráľa park in Bratislava (Slovakia)

Some early parks include the la Alameda de Hércules, in Seville, a promenaded public mall, urban garden and park built in 1574, within the historic center of Seville; the City Park, in Budapest, Hungary, which was property of the Batthyány family and was later made public.

An early purpose built public park was Derby Arboretum which was opened in 1840 by Joseph Strutt for the mill workers and people of the city. This was closely followed by Princes Park in the Liverpool suburb of Toxteth, laid out to the designs of Joseph Paxton from 1842 and opened in 1843. The land on which the Princes park was built was purchased by Richard Vaughan Yates, an iron merchant and philanthropist, in 1841 for £50,000. The creation of Princes Park showed great foresight and introduced a number of highly influential ideas. First and foremost was the provision of open space for the benefit of townspeople and local residents within an area that was being rapidly built up. Secondly it took the concept of the designed landscape as a setting for the suburban domicile, an idea pioneered by John Nash at Regent's Park, and re-fashioned it for the provincial town in a most original way. Nash's remodeling of St James's Park from 1827 and the sequence of processional routes he created to link The Mall with Regent's Park completely transformed the appearance of London's West End. With the establishment of Princes Park in 1842, Joseph Paxton did something similar for the benefit of a provincial town, albeit one of international stature by virtue of its flourishing mercantile contingent. Liverpool had a burgeoning presence on the scene of global maritime trade before 1800 and during the Victorian era its wealth rivalled that of London itself.

The form and layout of Paxton's ornamental grounds, structured about an informal lake within the confines of a serpentine carriageway, put in place the essential elements of his much imitated design for Birkenhead Park. The latter was commenced in 1843 with the help of public finance and deployed the ideas he pioneered at Princes Park on a more expansive scale. Frederick Law Olmsted visited Birkenhead Park in 1850 and praised its qualities. Indeed, Paxton is widely credited as having been one of the principal influences on Olmsted and Calvert's design for New York's Central Park of 1857.

Central Park in New York City is the most-visited urban park in the U.S.

Another early public park is the Peel Park, Salford, England opened on August 22, 1846. Another possible claimant for status as the world's first public park is Boston Common (Boston, Massachusetts, USA), set aside in 1634, whose first recreational promenade, Tremont Mall, dates from 1728. True park status for the entire common seems to have emerged no later than 1830, when the grazing of cows was ended and renaming the Common as Washington Park was proposed (renaming the bordering Sentry Street to Park Street in 1808 already acknowledged the reality).

Linear Parks

A linear park is a park that has a much greater length than width. A typical example of a linear park is a section of a former railway that has been converted into a park called a rail trail or greenway (i.e. the tracks removed, vegetation allowed to grow back). Parks are sometimes made out of oddly shaped areas of land, much like the vacant lots that often become city neighborhood parks. Linked parks may form a greenbelt.

Country Parks

In some countries, especially the United Kingdom, *country parks* are areas designated for recreation, and managed by local authorities. They are often located near urban populations, but they provide recreational facilities typical of the countryside rather than the town.

Private Parks

Private parks are owned by individuals or businesses and are used at the discretion of the owner. There are a few types of private parks, and some which once were privately maintained and used have now been made open to the public.

Hunting parks were originally areas maintained as open space where residences, industry and farming were not allowed, often originally so that nobility might have a place to hunt . These were known for instance, as *deer parks* (deer being originally a term meaning any wild animal). Manycountry houses in GreatBritain and Ireland still have parks of this sort, which since the 18th century have oftenbeen landscaped for aesthetic effect. They are usually a mixture of open grassland with scattered trees and sections of woodland, and are often enclosed by a high wall. Thearea immediately around the house is the garden. In some cases this will also featuresweeping lawns and scattered trees; the basic difference between a *country house's park* and its garden is that the park is grazed by animals, but they are excluded from the garden.

Shopping Mall

Westfield Garden State Plaza is an upscale shopping mall in Paramus, New Jersey, United States.

The interior of the Aviapark in Moscow, Russia.

The interior of the Toronto Eaton Centre in Toronto, Canada.

A shopping mall is a modern, chiefly North American, term for a form of shopping precinct or shopping center, in which one or more buildings form a complex of shops representing merchandisers with interconnecting walkways that enable customers to walk from unit to unit. A shopping arcade is a specific form serving the same purpose.

Many early shopping arcades such the Burlington Arcade in London, the Galleria Vittorio Emanuele II in Milan, and numerous arcades in Paris are famous and still trading. However, many smaller arcades have been demolished, replaced with large centers or "malls", often accessible by vehicle. Technical innovations such as electric lighting and escalators were introduced from the late 19th century.

From the late 20th century, entertainment venues such as movie theaters and restaurants began to be added. As a single built structure, early shopping centers were often architecturally significant constructions, enabling wealthier patrons to buy goods in spaces protected from the weather.

Regional Differences

Shopping arcade in Tokyo, Japan

In places around the world, the term *shopping centre* is used, especially in Europe, Australia, and South America. *Mall* is a term used predominantly in North America. Outside of North America, "shopping precinct" and "shopping arcade" are also used. In North America, Gulf countries, and India, the term *shopping mall* is usually applied to enclosed retail structures (and is generally abbreviated to simply *mall*), while *shopping centre* usually refers to open-air retail complexes; both types of facilities usually have large parking lots, face major traffic arterials, and have few pedestrian connections to surrounding neighbourhoods.

Robinsons Place Manila in Manila, Philippines

In the United Kingdom and Ireland, "malls" are commonly referred to as *shopping centres*. *Mall* primarily refers to either a shopping mall – a place where a collection of shops all adjoin a pedestrian area – or an exclusively pedestrianized street that allows shoppers to walk without interference from vehicle traffic. In North America, *mall* is generally used to refer to a large shopping area usually composed of a single building which contains multiple shops, usually "anchored" by one or more department stores surrounded by a parking lot, while the term "arcade" is more often used, especially in Britain, to refer to a narrow pedestrian-only street, often covered or between closely spaced buildings.

The majority of British shopping centres are located in city centres, usually found in old and historic shopping districts and surrounded by subsidiary open air shopping streets. Large examples include West Quay in Southampton; Manchester Arndale; Bullring Birmingham; Liverpool One; Buchanan Galleries in Glasgow; and Eldon Square in Newcastle upon Tyne. In addition to the inner city shopping centres, large UK conurbations will also have large out-of-town "regional malls" such as Meadowhall Centre, Sheffield serving South Yorkshire, the Trafford Centre in Greater Manchester and Bluewater in Kent. These centres were built in the 1980s and 1990s, but planning regulations prohibit the construction of any more. Out-of-town shopping developments in the UK are now focused on retail parks, which consist of groups of warehouse style shops with individual entrances from outdoors. Planning policy prioritizes the development of existing town centres, although with patchy success. Westfield Stratford City, in Stratford (London), is the largest shopping centre in Europe with over 330 shops, 50 restaurants and an 11 screen cinema and Westfield London is the largest inner-city shopping center in Europe. Bullring, Birmingham is the busiest shopping centre in the UK welcoming over 36.5 million shoppers in its opening year. There are a reported 222 malls in Europe. In 2014, these malls had combined sales of $12.47 billion. This represented a 10% bump in revenues from the prior year.

Development of Shopping Areas and Building Types

Headquarters of the International Council of Shopping Centers (in the middle skyscraper are the offices), the global trade association of the shopping center industry, in Midtown Manhattan, New York City.

Cabot Circus in Bristol city centre, England

One of the earliest examples of public shopping malls comes from ancient Rome, in forums where shopping markets were located. One of the earliest public shopping centers is Trajan's Market in Rome located in Trajan's Forum. Trajan's Market was probably built around 100-110 CE by Apollodorus of Damascus, and it is thought to be the world's oldest shopping center – a forerunner of today's shopping mall. The Grand Bazaar of Istanbul was built in the 15th century and is still one of the largest covered shopping centers in the world, with more than 58 streets and 4,000 shops. Numerous other covered shopping arcades, such as the 19th-century Al-Hamidiyah Souq in Damascus, Syria, might also be considered as precursors to the present-day shopping mall. Isfahan's Grand Bazaar, which is largely covered, dates from the 10th century. The 10-kilometer-long, covered Tehran's Grand Bazaar also has a lengthy history. The oldest continuously occupied shopping mall in the world is likely to be the Chester Rows. Dating back at least to the 13th century, these covered walkways housed shops, with storage and accommodation for traders on various levels. Different rows specialised in different goods, such as 'Bakers Row' or 'Fleshmongers Row'.

Gostiny Dvor in St. Petersburg, which opened in 1785, may be regarded as one of the first purposely-built mall-type shopping complexes, as it consisted of more than 100 shops covering an area of over 53,000 m² (570,000 sq ft).

The Marché des Enfants Rouges in Paris opened in 1628 and still runs today. The Oxford Covered Market in Oxford, England opened in 1774 and still runs today.

The Passage du Caire was opened in Paris in 1798. The Burlington Arcade in London was opened in 1819. The Arcade in Providence, Rhode Island introduced the retail arcade concept to the United States in 1828. The Galleria Vittorio Emanuele II in Milan, Italy followed in the 1870s and is closer to large modern malls in spaciousness. Other large cities created arcades and shopping centers in the late 19th century and early 20th century, including the Cleveland Arcade, Dayton Arcade and Moscow's GUM, which opened in 1890. Early shopping centers designed for the automobile include Market Square, Lake Forest, Illinois (1916), and Country Club Plaza, Kansas City, Missouri (1924).

Hongyuan outdoor mall in Shanghai

An early indoor mall prototype in the United States was the Lake View Store at Morgan Park, Duluth, Minnesota, which was built in 1915 and held its grand opening on July 20, 1916. The architect was Dean and Dean from Chicago and the building contractor was George H. Lounsberry from Duluth. The building is two stories with a full basement, and shops were originally located on all three levels. All of the stores were located within the interior of the mall; some shops were accessible from the inside and out.

In the mid-20th century, with the rise of the suburb and automobile culture in the United States, a new style of shopping center was created away from downtown. From early on, the design tended to be inward-facing, with malls following theories of how customers could best be enticed in a controlled environment. Similar, the concept of a mall having one or more "anchor stores" or "big box stores" was pioneered early, with individual stores or smaller-scale chain stores intended to benefit from the shoppers attracted by the big stores. Mall construction in America was encouraged by the accelerated depreciation laws of 1954, which incentivized greenfield development on the urban fringe. A second stimulus came from legislation passed in 1960, which allowed investors to band together in REITs (Real Estate Investment Trusts) to avoid corporate income taxes. The laws helped to shape the familiar exurban landscape of malls, motels, and fast food chains. By year end 2015, the total amount of retail space per person in the United States will be 48.3 square feet. This is off from the high of 49.8 square feet set in 2009. It is projected that this figure will continue to decline through 2020.

Great Western Arcade, Birmingham, England, built 1865

Early Examples

The Cleveland Arcade was among the first indoor shopping arcades in the US, and like its European counterparts, was an architectural triumph. When the building opened in 1890, two sides of the arcade had 1,600 panes of glass set in iron framing and is a prime example of Victorian architecture.

The early shopping center in the United States took shape at the Grandview Avenue Shopping Center (the "Bank Block") in Grandview Heights, Ohio in 1928, the first regional shopping center in America that integrated parking into the design. This general plan by Don Monroe Casto Sr. became the prototype of shopping centers for several decades. Other important shopping centers built in the 1920s and early 1930s include Country Club Plaza in Kansas City, Missouri, the Highland Park Village in Dallas, Texas; River Oaks in Houston, Texas; and the Park and Shop in Washington, D.C..

The suburban shopping center concept evolved further in the United States after World War II. Bellevue Shopping Square (now known as Bellevue Square) opened in 1946 in Bellevue, Washington, a suburb of Seattle. Town & Country Village also opened in 1946 in Sacramento, California. Then came the Broadway-Crenshaw Center (known today as Baldwin Hills Crenshaw Plaza), which was dedicated, in Los Angeles, in 1947. Two more suburban shopping centers were completed in 1949. Town and Country Drive-In Shopping Center (Town and Country Shopping Center), in Whitehall, Ohio was a strip-type complex erected in the environs of Columbus, Ohio. Park Forest, Illinois' Park Forest Plaza (Park Forest Downtown) was built along the lines of a cluster-type complex. It was situated in the southern suburbs of Chicago, Illinois.

The suburban shopping mall came into being with the opening of Seattle's Northgate Center (presently known as Northgate Mall) in April 1950. This was followed by Lakewood Center (1951), in Lakewood, California; Shoppers' World (1951), in Framingham, Massachusetts; Stonestown Center (now Stonestown Galleria) (1952) in San Francisco, California; and Northland Center (1954), in Southfield, Michigan. Open-air-type malls were also built in Canada and Australia. Don Mills Convenience Centre (now Shops at Don Mills) opened in 1955, in Toronto, Ontario. Chermside Drive-In Shopping Centre started trading to the public in 1957, in Brisbane, Australia.

The fully enclosed shopping mall did not appear until the mid-1950s. One of the earliest examples was the Valley Fair Shopping Center in Appleton, Wisconsin, which opened in March 1955. Valley Fair featured a number of modern features including a large parking area, anchor stores, and restaurants. The idea of a regionally-sized, fully enclosed shopping complex was pioneered in 1956 by the Austrian-born architect and American immigrant Victor Gruen. This new generation of regional-size shopping centers began with the Gruen-designed Southdale Center, which opened in the Twin Cities suburb of Edina, Minnesota, United States in October 1956. For pioneering the soon-to-be enormously popular mall concept in this form, Gruen has been called the "most influential architect of the twentieth century" by Malcolm Gladwell.

The first retail complex to be promoted as a "mall" was Paramus, New Jersey's Bergen Mall. The center, which opened with an open-air format in 1957, was enclosed in 1973. Aside from Southdale Center, significant early enclosed shopping malls were Harundale Mall (1958), in Glen Burnie, Maryland, Big Town Mall (1959), in Mesquite, Texas, Chris-Town Mall (1961), in Phoenix, Arizona, and Randhurst Center (1962), in Mount Prospect, Illinois.

The first fully enclosed shopping mall in Canada was Wellington Square. It was designed for Eaton's by John Graham, Jr. as an enclosed mall with a department store anchor and subterranean parking which opened in downtown London, Ontario, on August 11, 1960. After several renovations, it remains open today as Citi Plaza.

Other early malls moved retailing away from the dense, commercial downtowns into the largely residential suburbs. This formula (enclosed space with stores attached, away from downtown, and accessible only by automobile) became a popular way to build retail across the world. Gruen himself came to abhor this effect of his new design; he decried the creation of enormous "land wasting seas of parking" and the spread of suburban sprawl.

In the UK, Chrisp Street Market was the first pedestrian shopping area built with a road at the shop fronts. The first mall-type shopping precinct in Great Britain was built in the downtown area of Birmingham. Known as Bull Ring Centre (now Bull Ring Birmingham), it was officially dedicated in May 1964. A notable example is the Halton Lea Shopping Centre (originally known as Shopping City) in Runcorn, which opened in

1972 and was conceived as the center point for the new town's development. Another early example is the Brent Cross Centre, Britain's first out-of-town shopping mall and located on the northern outskirts of London, which was opened in March 1976.

In the United States, developers such as A. Alfred Taubman of Taubman Centers extended the concept further in 1980, with terrazzo tiles at the Mall at Short Hills in New Jersey, indoor fountains, and two levels allowing a shopper to make a circuit of all the stores. Taubman believed carpeting increased friction, slowing down customers, so it was removed. Fading daylight through glass panels was supplemented by gradually increased electric lighting, making it seem like the afternoon was lasting longer, which encouraged shoppers to linger.

Ala Moana Center in Honolulu, Hawaii is currently the largest open-air mall in the world and was one of the largest malls in the United States when it opened for business in August 1959. It is currently the sixteenth largest in the country. The Outlets at Bergen Town Center, the oldest enclosed mall in New Jersey, opened in Paramus on November 14, 1957, with Dave Garroway, host of *The Today Show*, serving as master of ceremonies. The mall, located just outside New York City, was planned in 1955 by Allied Stores to have 100 stores and 8,600 parking spaces in a 1,500,000 sq ft (140,000 m²) mall that would include a 300,000 sq ft (28,000 m²) Stern's store and two other 150,000 sq ft (14,000 m²) department stores as part of the design. Allied's chairman B. Earl Puckett confidently announced The Outlets at Bergen Town Center as the largest of ten proposed centers, stating that there were 25 cities that could support such centers and that no more than 50 malls of this type would ever be built nationwide.

In the 1970s, the Ontario government created the Ontario Downtown Renewal Programme, which helped finance the building of several downtown malls across Ontario such as Eaton Centre. The program was created to reverse the tide of small business leaving downtowns for larger sites surrounding the city. In the first quarter of 2012 shopping mall private investment hit an all-time low under 0.1 percent.

Largest Shopping Malls

Panoramic view of the SM Mall of Asia in the Philippines

The largest mall in the world is the New South China Mall in Dongguan, China with a gross floor area of 892,000 m² (9,600,000 sq ft). The world's second-largest shopping mall is the Golden Resources Mall in Beijing, China with a gross floor area of 680,000 m² (7,300,000 sq ft). SM Megamall in the Philippines, is the world's third-largest at 542,980 m² (5,844,600 sq ft) of gross floor area. The fourth largest shopping mall in the world is SM City North EDSA in Quezon City, Philippines with a gross floor area of 504,900 m² (5,435,000 sq ft) and the fifth largest shopping mall is 1 Utama in Malaysia at 465,000 m² (5,010,000 sq ft) of gross floor area.

The former Camp Snoopy Amusement park before it was Nickelodeon Universe at the center of the Mall of America in Bloomington, Minnesota, the largest shopping mall in the United States

Previously, the title of the largest enclosed shopping mall was with the West Edmonton Mall in Edmonton, Alberta, Canada from 1986 to 2004. It is now the fifth largest mall.

One of the world's largest shopping complexes in one location is the two-mall agglomeration of the Plaza at King of Prussia and the Court at King of Prussia in the Philadelphia suburb of King of Prussia, Pennsylvania, United States. The King of Prussia mall has the most shopping per square foot in the US.

The most visited shopping mall in the world and largest mall in the United States is the Mall of America, located near the Twin Cities in Bloomington, Minnesota. However, several Asian malls are advertised as having more visitors, including Mal Taman Anggrek, Kelapa Gading Mall and Pluit Village, all in Jakarta, Indonesia; Berjaya Times Square in Malaysia; SM City North EDSA, SM Mall of Asia and SM Megamall, all in Metro Manila, Philippines. The largest mall in South Asia is Bashundhara City in Dhaka, Bangladesh.

Types

The International Council of Shopping Centers classifies shopping malls into eight basic types: neighborhood center, community center, regional center, superregional center, fashion/specialty center, power center, theme/festival center, and outlet center. These definitions, published in 1999, were not restricted to shopping centers in any particular country, but later editions were made specific to the US with a separate set for Europe.

Neighborhood Center

A neighborhood center in the form of a strip mall, in Cornelius, Oregon

Neighborhood centers are small-scale malls serving the local neighborhood. They typically have a supermarket or a drugstore as an anchor, and are commonly arranged in a strip mall format. Neighborhood centers usually have a retail area of 30,000 to 150,000 square feet (2,800 to 13,900 m^2), and serve a primary area in a 3-mile (4.8 km) radius. They are sometimes known as convenience centers.

Community Mall

Community malls are larger than neighborhood centers, and offer a wider range of goods. They usually feature two anchor stores which are larger than that of a neighborhood center's, e.g. a discount department store. They may also follow a strip configuration, or may be L- or U-shaped. Community centers usually feature a retail area of 100,000 to 350,000 square feet (9,300 to 32,500 m^2) and serve a primary area of 3 to 6 miles (4.8 to 9.7 km).

Regional Center

A regional mall is, per the International Council of Shopping Centers, in the United States, a shopping mall which is designed to service a larger area than a conventional shopping mall. As such, it is typically larger with 400,000 sq ft (37,000 m^2) to 800,000 sq ft (74,000 m^2) gross leasable area with at least two anchor stores and offers a wider selection of stores. Given their wider service area, these malls tend to have higher-end stores that need a larger area in order for their services to be profitable but may have discount department stores. Regional malls are also found as tourist attractions in vacation areas.

The Lifestyle Center La Gran Via mall in San Salvador City, El Salvador.

Superregional Center

SM Southmall, a superregional mall in Metro Manila, Philippines, with a gross leasable area of 2,207,900 sq ft (205,120 m²)

A super regional mall is, per the International Council of Shopping Centers, in the US a shopping mall with over 800,000 sq ft (74,000 m²) of gross leasable area, three or more anchors, mass merchant, more variety, fashion apparel, and serves as the dominant shopping venue for the region (25 miles) in which it is located.

Westfield Carousel, in a suburb of Perth, Australia, with a gross leasable area of 82,874 m²
(892,050 sq ft)

Fashion/Specialty Center

Fashion or specialty centers feature upscale apparel shops and boutiques and cater to customers with higher incomes. They usually have a retail area ranging from 80,000 to 250,000 square feet (7,400 to 23,200 m²) and serve an area of 5 to 15 miles (8.0 to 24.1 km).

Power Center

Power centers are large shopping centers that almost exclusively feature several big-box retailers as their anchors. They usually have a retail area of 250,000 to 600,000 square feet (23,000 to 56,000 m²) and a primary trade area of 5 to 10 miles (8.0 to 16.1 km).

Theme/Festival Center

Theme or festival centers have distinct unifying themes that are followed by their individual shops as well as their architecture. They are usually located in urban areas and cater to tourists. They typically feature a retail area of 80,000 to 250,000 square feet (7,400 to 23,200 m²).

Terminal 21 in Bangkok has an airport/world cities theme, reflected in this escalator sign.

Outlet Center

An outlet mall (or outlet center) is a type of shopping mall in which manufacturers sell their products directly to the public through their own stores. Other stores in outlet malls are operated by retailers selling returned goods and discontinued products, often at heavily reduced prices. Outlet stores were found as early as 1936, but the first multi-store outlet mall, Vanity Fair, located in Reading, PA did not open until 1974. Belz Enterprises opened the first enclosed factory outlet mall in 1979, in Lakeland, TN, a suburb of Memphis.

Components

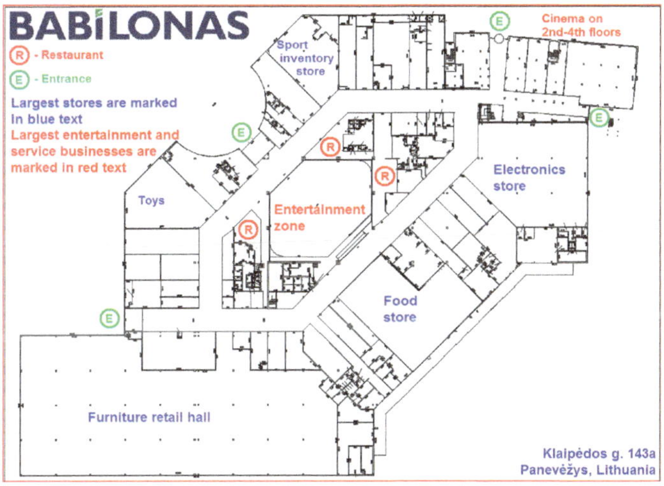

The layout of a mid-sized shopping center Babilonas in Panevėžys, Lithuania (with main stores marked in text).

Food Court

A common feature of shopping malls is a food court: this typically consists of a number of fast food vendors of various types, surrounding a shared seating area.

Department Stores

When the shopping mall format was developed by Victor Gruen in the mid-1950s, signing larger department stores was necessary for the financial stability of the projects, and to draw retail traffic that would result in visits to the smaller stores in the mall as well. These larger stores are termed anchor store or draw tenant. In physical configuration, anchor stores are normally located as far from each other as possible to maximize the amount of traffic from one anchor to another.

Stand-Alone Stores

Frequently, a shopping mall or shopping center will have satellite buildings located either on the same tract of land or on one abutting it, on which will be located stand-alone stores, which may or may not be legally connected to the central facility through contract or ownership. These stores may have their own parking lots, or their lots may interconnect with those of the mall or center. The existence of the stand-alone store may have been planned by the mall's developer, or may have come about through opportunistic actions by others, but visually the central facility – the mall or shopping center – and the satellite buildings will often be perceived as being a single "unit", even in circumstances where the outlying buildings are not officially or legally connected to the mall in any way.

Dead Malls

Belz Factory Outlet Mall, an abandoned shopping mall in Allen, Texas, United States

In the United States, in the mid-1990s, malls were still being constructed at a rate of 140 a year. But in 2001, a PricewaterhouseCoopers study found that underperforming and vacant malls, known as "greyfield" and "dead mall" estates, were an emerging problem. In 2007, a year before the Great Recession, no new malls were built in America, for the first time in 50 years. City Creek Center Mall in Salt Lake City, which opened in March 2012, was the first to be built since the recession.

In recent years, the number of dead malls increased significantly in the early twenty first century because the economic health of malls across the United States has been in decline, with high vacancy rates in these malls. From 2006 to 2010, the percentage of malls that are considered to be "dying" by real estate experts (have a vacancy rate of at least 40%), unhealthy (20–40%), or in trouble (10–20%) all increased greatly, and these high vacancy rates only partially decreased from 2010 to 2014. In 2014, nearly 3% of all malls in the United States were considered to be "dying" (40% or higher vacancy rates) and nearly one-fifth of all malls had vacancy rates considered "troubling" (10% or higher). Some real estate experts say the "fundamental problem" is a glut of malls in many parts of the country creating a market that is "extremely over-retailed".

New Trends

In parts of Canada, it is now rare for new shopping malls to be built. The Vaughan Mills Shopping Centre, opened in 2004, and Crossiron Mills, opened in 2009, are the only malls built in Canada since 1992. Outdoor outlet malls or big box shopping areas known as power centers are now favored, although the traditional enclosed shopping mall is still in demand by those seeking weather-protected, all-under-one-roof shopping. In addition, the enclosed interconnections between downtown multi story shopping malls continue to grow in the Underground city of Montreal (32 kilometres of passageway), the PATH system of Toronto (27 km (17 mi) of passageway) and the Plus15 system of Calgary (16 km (9.9 mi) of overhead passageway).

In Russia, on the other hand, as of 2013 a large number of new malls had been built near major cities, notably the MEGA malls such as Mega Belaya Dacha mall near Moscow. In large part they were financed by international investors and were popular with shoppers from the emerging middle class.

In the United States, owners are making drastic moves to convert struggling malls. This includes converting malls into apartments, offices and industrial space. Other owners have taken the approach to turning large chunks of malls into parks and playgrounds. In Austin, Texas, Highland Mall is in the midst of a conversion to a community college. Much of the 600,000 square foot mall will be the home of Austin Community College.

Vertical Malls

High land prices in populous cities have led to the concept of the "vertical mall," in which space allocated to retail is configured over a number of stories accessible by elevators and/or escalators (usually both) linking the different levels of the mall. The challenge of this type of mall is to overcome the natural tendency of shoppers to move horizontally and encourage shoppers to move upwards and downwards. The concept of a vertical mall was originally conceived in the late 1960s by the Mafco Company, former shopping center development division of Marshall Field & Co. The Water Tower Place skyscraper, Chicago, Illinois, was built in 1975 by Urban Retail Properties. It contains a hotel, luxury condominiums, and office space and sits atop a block-long base containing an eight-level atrium-style retail mall that fronts on the Magnificent Mile.

The Fashion Centre at Pentagon City, in Arlington, Virginia, United States

Vertical malls are common in densely populated conurbations such as Hong Kong and Bangkok. Times Square in Hong Kong is a principal example.

A vertical mall may also be built where the geography prevents building outward or there are other restrictions on construction, such as historical buildings or significant archeology. The Darwin Shopping Centre and associated malls in Shrewsbury, UK, are built on the side of a steep hill, around the former outer walls of the nearby medieval castle; consequently the shopping center is split over seven floors vertically – two locations horizontally – connected by elevators, escalators and bridge walkways. Some establishments incorporate such designs into their layout, such as Shrewsbury's McDonalds restaurant, split into four stories with multiple mezzanines which feature medieval castle vaults – complete with arrowslits – in the basement dining rooms.

Online Shopping Influence

Faced with the exploding popularity of buying online, shopping malls are emptying and are seeking new solutions to generate traffic. In the US, for example, roughly 200 out of 1,300 malls across the country are going out of business. To combat this trend, developers are trying to turn malls into leisure centers that include attractions such as parks, movie theaters, gyms, and even fishing lakes. Others, such as the European commercial real-estate giant Unibail-Rodamco, are modernizing their approach by promoting brand interaction and enhanced architectural appeal. A recent example that integrates both approaches is the So Ouest mall outside of Paris that was designed to resemble elegant, Louis XV-style apartments and includes 17,000 square metres (180,000 sq ft) of green space. The Australian mall company Westfield launched an online mall (and later a mobile app) with 150 stores, 3,000 brands and over 1 million products. Online shopping has increased its share of total retail sales since 2008. In Q3 2008, it comprised 3.6% of retail purchases and this increased to 7.4% by Q3 2015.

Shopping Property Management Firms

A shopping property management firm is a company that specializes in owning and managing shopping malls. Most shopping property management firms own at least 20 malls. Some firms use a similar naming scheme for most of their malls; for example, Mills Corporation puts "Mills" in most of its mall names and SM Prime Holdings of the Philippines puts "SM" in all of its malls, as well as anchor stores such as SM Department Store, SM Appliance Center, SM Hypermarket, SM Cinema, and SM Supermarket. In the UK, The Mall Fund changes the name of any center it buys to *"The Mall (location)"*, using its pink-M logo; when it sells a mall the center reverts to its own name and branding, such as the Ashley Centre in Epsom.

Shopping center management and advisory firms are bringing about professional management practices to the largely fragmented shopping center development industry in India. Historically, land ownership in India, has been fragmented and as a byproduct shopping center development, which rendered the single mall developers vulnerable to dubious advice and practices, since standard benchmarks, knowledge resources, and skilled people were scarce. This is changing as new firms promoted by former shopping center managers are stepping in to bridge the gap between ownership and professional management.

Beyond Squarefeet from India is another mall management company, which is foraying into various other countries such as India, Iran, Nepal, Nigeria, Qatar, etc. Mall management is slowly becoming a trend and is much sought after services in Asia and other markets.

Legal Issues

One controversial aspect of malls has been their effective displacement of traditional main streets or high streets. Some consumers prefer malls, with their parking garages, controlled environments, and private security guards, over CBDs or downtowns, which frequently have limited parking, poor maintenance, outdoor weather, and limited police coverage.

In response, a few jurisdictions, notably California, have expanded the right of freedom of speech to ensure that speakers will be able to reach consumers who prefer to shop, eat, and socialize within the boundaries of privately owned malls.

Gallery

World's Largest Shopping Malls/Centers

Lulu Mall-Cochin_ The largest shopping mall in India.

Isfahan City Center a large shopping mall in Isfahan, Iran, Interior.

The Port Grand Food and Entertainment Complex in Karachi is not only one of the most high-profile shopping mall in the city, but is also Asia's largest food street.

Berjaya Times Square in Kuala Lumpur, Malaysia

SM City North EDSA in Quezon City, Philippines

1 Utama in Petaling Jaya, Malaysia

Abasto de Buenos Aires in Buenos Aires, Argentina

Plaza Las Américas, the largest shopping mall in the Caribbean in San Juan, Puerto Rico

The Dubai Mall in Dubai, UAE

Canal Walk in Cape Town, South Africa

Galerías Pacífico in Buenos Aires, Argentina

Tunjungan Plaza in Surabaya, Indonesia

CentralWorld in Bangkok, Thailand

West Edmonton Mall in Alberta, Canada

Iguatemi Fortaleza Shopping in Fortaleza, Brazil

Mal Taman Anggrek in Jakarta, Indonesia

Sunway Pyramid in Subang Jaya, Malaysia

Siam Paragon in Bangkok, Thailand

The Village Mall in Rio de Janeiro, Brazil

Gateway Theatre of Shopping in Durban, South Africa

References

- Davreu, Robert (1978). "Cities of Mystery: The Lost Empire of the Indus Valley". *The World's Last Mysteries.* (second edition). Sydney: Readers' Digest. pp. 121-129. ISBN 0-909486-61-1.
- *Boerefijn, Wim (2010). The foundation, planning and building of new towns in the 13th and 14th centuries in Europe. An architectural-historical research into urban form and its creation. Phd. thesis Universiteit van Amsterdam. ISBN 978-90-9025157-8.*
- Smith Morris et al. *British Town Planning and Urban Design*, 1997, ISBN 0-582-23496-4, Longman, Singapore.
- *Gunawan Tjahjono, ed. (1998). Architecture. Indonesian Heritage 6. Singapore: Archipelago Press. p. 109. ISBN 981-3018-30-5.*
- *Dino Fanara (2006). Angel of the East Indies: Biography of the Van Dooremolen Family. iUniverse. p. 55. ISBN 9780595860449.*
- *Silver, Christopher (2007). Planning the Megacity: Jakarta in the Twentieth Century - Planning, History and Environment Series. Routledge. p. 101. ISBN 9781135991227.*
- Boerefijn, Wim (2010) "The foundation, planning and building of new towns in the 13th and 14th centuries in Europe. An architectural-historical research into urban form and its creation". Phd. thesis Universiteit van Amsterdam. ISBN 978-90-9025157-8.
- *Geist, Johann Friedrich. Arcades: The History of a Building Type. Cambridge, Mass: MIT Press*

(MA). ISBN 0262570629.

- *MacKeith, Margaret (1986). The history and conservation of shopping arcades. London, New York: Mansell Pub. ISBN 0720117577.*
- *Urban Geography: A Global Perspective* Michael Pacione, (Routledge, Informa UK Ltd. 2001) ISBN 978-0-415-19195-1.
- *Bathroom Reader's Institute. "A History of the Shopping Mall, Part III". Uncle John's Heavy Duty Bathroom Reader. Bathroom Reader's Press. p. 401. ISBN 978-1-60710-183-3*
- *O'Donahue, Tony (May 1, 2005). The Tale of a City: Re-Engineering the Urban Environment. Toronto: Dundurn Press Ltd. p. 43. ISBN 978-1550025569. (subscription required (help)).*
- *Frieden, Bernard J.; Sagalyn, Lynne B. (1989). Downtown, Inc.: How America Rebuilds Cities. Cambridge, MA: MIT Press. p. 233. ISBN 978-0262560597. Retrieved 2015-10-21.*
- Wilson, Thomas D. *The Ashley Cooper Plan: The Founding of Carolina and the Origins of Southern Political Culture*. Chapel Hill, NC: University of North Carolina Press, 2016. pp. 12-13.
- *Agovino, Theresa. "Investors Find Bargains at Europe's Outlet Malls". Wall Street Journal. ISSN 0099-9660. Retrieved 2015-12-03.*
- *Pleven, Liam (November 24, 2015). "Shrinking U.S. Shopping Malls Get Makeovers". Wall Street Journal. ISSN 0099-9660. Retrieved November 30, 2015.*
- *"US Shopping-Center Classification and Characteristics" (PDF). International Council of Shopping Centers. August 2015. Retrieved November 13, 2015.*
- *Millar, Lisa (January 20, 2015). "Dead malls: Half of America's shopping centres predicted to close by 2030". Australian Broadcasting Corporation News. Retrieved 2015-10-21.*
- *Schwartz, Nelson D. (January 3, 2015). "The Economics (and Nostalgia) of Dead Malls". The New York Times. Retrieved 2015-10-21.*
- *"Calgarys Plus15 Skywalk". City of Calgary. Archived from the original on December 25, 2014. Retrieved 2015-10-21.*
- *"Westfield launches new 'online mall' with 100,000 products". The Australian Financial Review. November 19, 2013. Retrieved 2015-10-21.*

Permissions

All chapters in this book are published with permission under the Creative Commons Attribution Share Alike License or equivalent. Every chapter published in this book has been scrutinized by our experts. Their significance has been extensively debated. The topics covered herein carry significant information for a comprehensive understanding. They may even be implemented as practical applications or may be referred to as a beginning point for further studies.

We would like to thank the editorial team for lending their expertise to make the book truly unique. They have played a crucial role in the development of this book. Without their invaluable contributions this book wouldn't have been possible. They have made vital efforts to compile up to date information on the varied aspects of this subject to make this book a valuable addition to the collection of many professionals and students.

This book was conceptualized with the vision of imparting up-to-date and integrated information in this field. To ensure the same, a matchless editorial board was set up. Every individual on the board went through rigorous rounds of assessment to prove their worth. After which they invested a large part of their time researching and compiling the most relevant data for our readers.

The editorial board has been involved in producing this book since its inception. They have spent rigorous hours researching and exploring the diverse topics which have resulted in the successful publishing of this book. They have passed on their knowledge of decades through this book. To expedite this challenging task, the publisher supported the team at every step. A small team of assistant editors was also appointed to further simplify the editing procedure and attain best results for the readers.

Apart from the editorial board, the designing team has also invested a significant amount of their time in understanding the subject and creating the most relevant covers. They scrutinized every image to scout for the most suitable representation of the subject and create an appropriate cover for the book.

The publishing team has been an ardent support to the editorial, designing and production team. Their endless efforts to recruit the best for this project, has resulted in the accomplishment of this book. They are a veteran in the field of academics and their pool of knowledge is as vast as their experience in printing. Their expertise and guidance has proved useful at every step. Their uncompromising quality standards have made this book an exceptional effort. Their encouragement from time to time has been an inspiration for everyone.

The publisher and the editorial board hope that this book will prove to be a valuable piece of knowledge for students, practitioners and scholars across the globe.

Index